Springer Theses

Recognizing Outstanding Ph.D. Research

For further volumes:
http://www.springer.com/series/8790

Aims and Scope

The series "Springer Theses" brings together a selection of the very best Ph.D. theses from around the world and across the physical sciences. Nominated and endorsed by two recognized specialists, each published volume has been selected for its scientific excellence and the high impact of its contents for the pertinent field of research. For greater accessibility to non-specialists, the published versions include an extended introduction, as well as a foreword by the student's supervisor explaining the special relevance of the work for the field. As a whole, the series will provide a valuable resource both for newcomers to the research fields described, and for other scientists seeking detailed background information on special questions. Finally, it provides an accredited documentation of the valuable contributions made by today's younger generation of scientists.

Theses are accepted into the series by invited nominated only and must fulfill all of the following criteria

- They must be written in good English.
- The topic of should fall within the confines of Chemistry, Physics and related interdisciplinary fields such as Materials, Nanoscience, Chemical Engineering, Complex Systems and Biophysics.
- The work reported in the thesis must represent a significant scientific advance.
- If the thesis includes previously published material, permission to reproduce this must be gained from the respective copyright holder.
- They must have been examined and passed during the 12 months prior to nomination.
- Each thesis should include a foreword by the supervisor outlining the significance of its content.
- The theses should have a clearly defined structure including and introduction accessible to scientists not expert in that particular field.

Symeon Karagiannidis

Catalytic Microreactors for Portable Power Generation

Doctoral Thesis accepted by
Swiss Federal Institute of Technology Zurich, Switzerland

Springer

Author
Dr. Symeon Karagiannidis
Paul Scherrer Institute and Swiss Federal Institute of Technology, Zurich
CH-5232 Villigen PSI
Switzerland
e-mail: s.karagiannidis@gmail.com

Supervisor
Dr. Ioannis Mantzaras
Paul Scherrer Institute and Swiss Federal Institute of Technology, Zurich
CH-5232 Villigen PSI
Switzerland
e-mail: ioannis.mantzaras@psi.ch

ISSN 2190-5053 e-ISSN 2190-5061

ISBN 978-3-642-17667-8 e-ISBN 978-3-642-17668-5

DOI 10.1007/978-3-642-17668-5

Springer Heidelberg Dordrecht London New York

Cover design: eStudio Calamar, Berlin/Figueres

Printed on acid-free paper

Springer is part of Springer Science+Business Media (www.springer.com)

Parts of this thesis have been published in the following journal articles

S. Karagiannidis, J. Mantzaras, "Numerical investigation on the start-up of methane-fueled catalytic microreactors", *Combustion and Flame*, **157**:1400–1413, 2010

S. Karagiannidis, J. Mantzaras, K. Boulouchos, "Stability of hetero-/homogeneous combustion in propane- and methane-fueled catalytic microreactors: Channel confinement and molecular transport effects", *Proceedings Combustion Institute*, **33**:3241–3249, 2011

S. Karagiannidis, K. Marketos, J. Mantzaras, R. Schaeren, K. Boulouchos, "Experimental and numerical investigation of a propane-fueled, catalytic mesoscale combustor", *Catalysis Today*, **155**:108–115, 2010

S. Karagiannidis, J. Mantzaras, R. Bombach, S. Schenker, K. Boulouchos, "Experimental and numerical investigation of the hetero-/homogenous combustion of lean propane/air mixtures over platinum", *Proceedings Combustion Institute*, **32**: 1947–1955, 2009

S. Karagiannidis, J. Mantzaras, G. Jackson, K. Boulouchos, "Hetero-/homogeneous combustion and stability maps in methane-fueled catalytic microreactors", *Proceedings Combustion Institute*, **31**:3309–3317, 2007 (Distinguished Paper Award)

pathway previously neglected in numerical studies of catalytic microreactors) is important in determining combustion stability even at high microreactor confinements, the substantial broadening of the stable combustion regime for hydrocarbon-fueled microcombustors with rising pressure, the dual role of surface radiation heat transfer at both steady-state operation and during the transient heat-up phase, as well as the counter-intuitive result that methane-fueled catalytic microreactors are more stable during high-inlet-velocity operation compared to propane-fueled ones, owing to methane's lower Lewis number and despite the higher catalytic and gas-phase reactivities of propane. In conclusion, the PhD thesis of Dr. Karagiannidis has provided valuable information with a high scientific impact on topics related to the design and operation of catalytic microreactors for portable power generation applications.

Villigen PSI, December 2010

PD Dr. Ioannis Mantzaras
Paul Scherrer Institute

Acknowledgments

This work was conducted in the Combustion Fundamentals Group of Paul Scherrer Institute, Villigen, Switzerland. Additional financial support was provided by the INIT Project "Ultra-High-Energy-Density Converters for Portable Power" of the Swiss Federal Institute of Technology Zurich.

First of all, I am grateful to Prof. Konstantinos Boulouchos, head of the Aerothermochemistry and Combustion Systems Laboratory, for giving me the opportunity to go through and complete my doctoral studies under his supervision.

I would also like to express my gratitude to my direct supervisor Dr. Ioannis Mantzaras, head of the Combustion Fundamentals Group, for his invaluable advice, support and input through my entire work.

Special thanks to Dr. Rolf Bombach for his knowledge and support in the laser diagnostics as well as to Rolf Schaeren for his work and assistance in the test rig; their input was critical in the realization of the experiments.

Many thanks to all former and current members of the Combustion Fundamentals Group in PSI, Salvatore Arcidiacono, Marta Bruska, Yohannes Ghermay, Gianmarco Pizza, Nikolaos Prasianakis, Adrian Schneider, for a fruitful and pleasant cooperation during my employment in Paul Scherrer Institute.

Last but not least, I would like to thank my entire family who supported me by all means possible through my studies, and my friends to which I owe a pleasant stay in Zurich.

Contents

List of Abbreviations

Nomenclature

B	Active-to-geometrical surface ratio
b	Channel half-height
c_p, c_s	Specific heat of gas at constant pressure, specific heat of solid
C_k	Concentration of k-th species
D_{km}	Mixture-average species diffusion coefficient
$D^T{}_k$	Species thermal diffusion coefficient
$F_{k\text{–}j}$	Radiation configuration factor between surface elements k and j
$\dot{H}_g$	Integral enthalpy flux in the gas-phase
h, h_k^o	Total enthalpy, chemical enthalpy of k-th gaseous species
k, k_s	Thermal conductivity of gas, thermal conductivity of solid
K_g	Total number of gaseous species
M_s	Total number of surface species
m	Mass flowrate
L, L_a	Total channel/reactor length, coated channel length
p	Pressure
q	Radiative flux, radiative flux on k-th discrete surface element, heat source term
R	Universal gas constant
r	Channel radius
$\dot{s}_k$	Heterogeneous molar production rate of k-th species
S/V	Surface-to-volume ratio
t	Time
t_{ig}, t_{st}	Ignition time, steady-state time
T, T_o	Temperature, reference temperature
u, U_{IN}	Streamwise velocity component, inlet streamwise velocity
v	Transverse velocity component
$\vec{V}_k$	Species diffusion velocity vector

W_k, $\bar{W}$	Gas-phase species molecular weight, mixture average molecular weight
X_k	Mole fraction of k-th gaseous species
Y_k	Mass fraction of k-th gaseous species
x, y	Streamwise and transverse coordinates

Greek Symbols

α, α_s	Gas thermal diffusivity, solid thermal diffusivity
Γ	Surface site density
δ	Channel wall thickness
ε	Surface emissivity
η	Efficiency
μ	Viscosity
ρ, ρs	Density of gas, density of solid
σ	Stefan-Boltzmann constant, monolith channel density
σ_m	Surface species site occupancy
$\tau_{CH,\ k}$	Characteristic chemical times of k-th species
χ	Homogeneous ignition streamwise distance
$\dot{\omega}_k$	Homogeneous molar production rate of k-th species

Subscripts

AD	Adiabatic
c	Combustor monolith
e	Electric
g	Gas
h	Hydraulic
IN, OUT	Inlet, outlet
k, m	Indices for gas-phase and surface species
r	Reactor
s	Solid
th	Thermal
W	Wall
x	Axial direction
y	Transverse direction
z	Lateral direction

Chapter 1
Introduction

1.1 Portable Power Generation: Motivation

In the last two decades significant advances have been achieved in the miniaturization and compacting of various electromechanical and electronic consumer devices such as laptop computers, mobile phones or even personal transport vehicles. Moreover, new needs have appeared in such diverse areas as medicine (e.g. artificial organs) and military applications (e.g. unmanned aerial vehicles) subsequently giving rise to new technological challenges [1–3]. The push towards even greater portability, integration and autonomous operation of such devices has given rise to new and promising research fields, such as the field of micro electro-mechanical systems (MEMS) [4]. Once the major constrains of precision fabrication in the microscale had been removed [5], attention was drawn to the main factor bottlenecking further advancements in scaling down such devices: the lack of efficient and reliable power sources.

The wide range of applications where portable power (aka 'palm-power') is required induces an associated large span of power and energy density requirements which small-scale power generation and storage units need to meet. Characteristically, the required power densities vary from 10 to 1,000 W/kg and energy densities from 500 to 5,000 Wh/kg [1]. While the upper power and energy density limits are not necessarily applicable to microscale systems, they nevertheless refer to power devices sufficiently small (length scales from 1 mm to 10 cm). Discerning the vast possibilities and gains from developing the technologies capable of meeting the present and future demands for on-the-go power generation, significant research efforts were spurred by various funding sources [6, 7], which succeeded in advancing the front of micro- and mesoscale power generation.

A common denominator of many related studies on portable power units has been the conclusion that, despite the latest achievements in electrochemical power

S. Karagiannidis, *Catalytic Microreactors for Portable Power Generation*,
Springer Theses, DOI: 10.1007/978-3-642-17668-5_1,

generation and storage [8], even state-of-the-art battery technology (e.g. Li-ion batteries) will be unable to deliver the levels of power and energy densities required for future portable electronic/electromechanical applications and autonomous operating devices [1]. At the same time, a number of significant theoretical and experimental works have exemplified and demonstrated the benefits of scaled-down power generating devices employing combustion-driven thermal cycles [2], with the use of either internal [9] or external [10] micro- and mesoscale combustion engines. The general notion behind such small-scale, combustion-driven engines is that the use of hydrocarbon fuel can, thanks to its high energy content per unit mass, render even relatively inefficient systems capable of delivering power densities almost an order of magnitude higher than those currently available from Li-ion batteries (see Fig. 1.1).

Various configurations of small-scale internal combustion engines have been employed towards the realization of a portable, hydrocarbon-fueled power generation system. Selected cases which have been investigated experimentally include a microscale Wankel-type rotary engine [11], a mesoscale unit employing a Stirling cycle [10] and a coupled thermoelectric/microreactor system [12]. One of the most promising research areas in microscale thermal engines is the micro-gas-turbine concept, essentially the pursuit of down-sizing conventional large-scale, gas-turbine-driven power plants, so as to deliver power outputs ranging from a few Watts up to 1 kW_e. In the standard approach followed, a single-stage compressor is coupled to a single-stage turbine on a common shaft with an appropriately sized generator, with a microburner affixed after the compressor discharge to provide the thermal power needed for the thermal cycle.

An early attempt towards the materialization of a micro-gas-turbine has been the MIT Microengine Project [9, 13], wherein a MEMS-based unit employing a

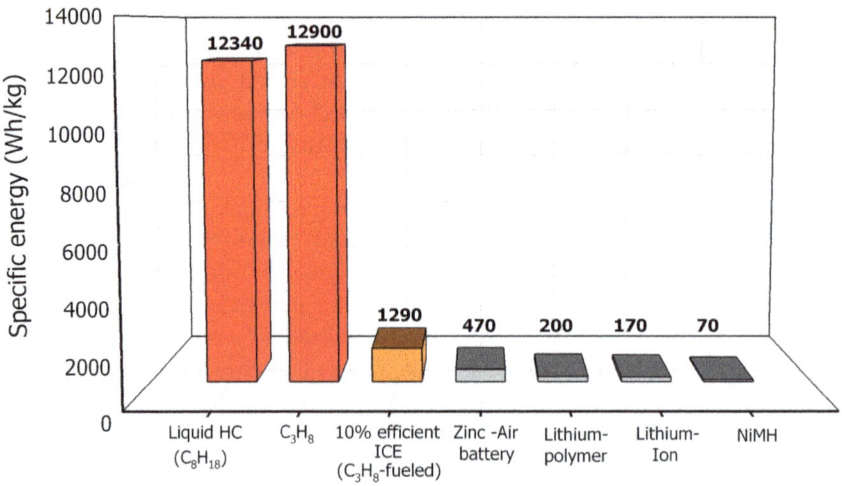

Fig. 1.1 Specific energy for selected hydrocarbon fuels, an internal combustion engine with a net efficiency of 10% and a number of representative battery types [2]

two-dimensional compressor/combustor/turbine configuration was used to convert hydrogen fuel in electrical power of the order of 10 W. Due to substantial technical difficulties in maintaining structural integrity and minimizing heat dissipation through the various system components, the aforementioned project has failed so far to produce a unit with positive power output. Nevertheless, this did not slow down the efforts towards a micro-gas-turbine-based portable unit (albeit with a number of compromises), due to the recognized potential of turbomachinery in delivering very high power densities. Realizing that a target power output of at least an order of magnitude higher than the MIT project would lift many of the technical obstacles inherent at the microscale, a number of research groups focused on the design of single-stage, micro-gas-turbine systems with a net power output of $\sim 100\ W_e$. Attempts at this up-scale level proved more successful, since the feasibility of the proposed systems has so far been proven at least at the level of individual components [14, 15].

A similar micro-gas-turbine project was recently initiated at the Swiss Federal Institute of Technology, based on a simple recuperated thermal cycle [16]. A schematic of the proposed design is presented in Fig. 1.2. Using a single-stage radial compressor and axial turbine coupled to a high-speed generator, a target power output of 100 W_e was set, with propane being the fuel of choice. By aiming for an overall system efficiency of $\sim 10\%$, the power output required from the combustor component was of the order of 1,000 W_{th}. This requirement, in conjunction with strict limitations imposed to the overall system size (each individual component having characteristic length scales no more than a few centimeters), gave rise to the need of designing a microscale burner capable of sustaining combustion with the highest possible efficiency under the operating conditions dictated by the micro-gas-turbine system design. Due to the associated large surface-to-volume (*S*/*V*) ratio of such micro- and mesoscale systems, a catalytic microreactor concept was selected as the most appropriate combustor solution.

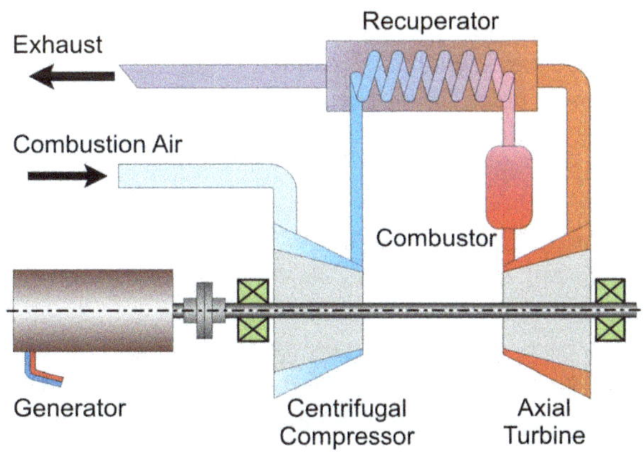

Fig. 1.2 Proposed micro-gas-turbine power generation unit [16]

1.2 Catalytic and Gas-Phase Microscale Combustion

Hydrocarbon- and hydrogen-fueled microreactors have received increased attention as energy sources for electrical and thermal power generation thanks to their superior energy density compared to that of state-of-the-art lithium batteries [2]. Although large-scale combustor technology cannot directly be applied on small-scale devices, pure gas-phase combustion of methane has been demonstrated at the sub-millimeter scale in reactors with appropriately annealed walls [17]. Moreover, practical configurations such as the mesoscale heat-recirculating "Swiss-roll" burner have been shown to stabilize near-stoichiometric propane/air flames for small channel widths of 3.5 mm [18]. Despite the feasibility of micro- and meso-scale gaseous combustors, the associated large surface-to-volume ratios at the microscale give rise to flame stabilization difficulties due to thermal and radical quenching on the combustor walls [19]. Catalytic combustion is a plausible solution to overcome these issues [20], since it satisfies the requirements of large *S*/*V* ratios and moderate operational temperatures relevant to micro-/meso-scale devices. Noble metal catalysts are employed in most cases, due to their superior activity for the total oxidation and reforming of hydrocarbons at modest reactor temperatures [21], the preferred catalyst in most cases being platinum. Heterogeneous fuel conversion was initially pursued in large power generation systems as an ultra-low-NO_x technology with enhanced flame stability under either fuel-lean or fuel-rich operating conditions [22–25], while methane, the main constituent of natural gas, has been the fuel of interest in the aforementioned studies.

A number of experimental studies have demonstrated the viability of small-scale combustion using different approaches, such as the heat-recirculating Swiss-roll burner which makes use of the excess exhaust gas enthalpy for increased combustion stability [18, 26], pure gas-phase combustors [27], as well as a variety of catalytic reactors such as catalytic foam structures [28] and conventional straight catalytic channels [29]. Additional experimental works have focused on hydrocarbon catalytic combustion specifically for microscale power generation [30–32]. The introduced concepts for catalytic microburners have been investigated in terms of combustion stability and efficiency, as well as microreactor thermal management, using a wide variety of fuels ranging from hydrogen [33] and methane [34] to the more practical for portable applications propane [35] and JP8 fuel [36]. Various other applications for catalytic microreactors range from catalytic microthrusters for space applications [37] to microreactors used for fuel reforming in micro solid oxide fuel cells [38].

As microscales typically hinder detailed measurements, there is a need of advanced numerical models for the simulation of both catalytic (heterogeneous) and gas-phase (homogeneous) microreactors. A number of computational tools has thus been developed, ranging from simple steady models for the performance assessment of methane-fueled catalytic microreactors without the inclusion of axial heat conduction in the channel walls [39], to transient models for the mapping of complex hydrogen/air flame dynamics in non-catalytic microchannels [40]. For catalytic microreactors, in particular, a number of studies have focused on key thermal management issues such as heat recirculation through the reactor walls [18], as well as reactor sizing and material properties [41]. Numerical simulations of catalytic microreactors found in the literature, which encompass investigations of combustion stability and performance, employ a diverse set of computational tools such as 1-D models with lumped heat and mass transport coefficients [41, 42], 2-D models with detailed chemistry but without heat conduction in the solid wall [39] and 2-D models with simplified chemistry accounting for heat conduction in the solid [43, 44]. Despite the large number of numerical models available in the literature for catalytic microreactors, no full 2-D CFD model with all relevant heat transfer mechanisms in the solid (including surface radiation heat transfer) has been reported; a lack of models employing both detailed catalytic and gas-phase chemistries is also evident. It should also be emphasized that although correlations for lumped transport coefficients (Nusselt and Sherwood numbers) have been proposed for laminar channel flows with catalytic reactions [45], their universal applicability under different reacting conditions is not warranted. Moreover, 1-D models fail to describe homogeneous combustion due to the strong dependence of gas-phase reactions on the boundary layer profiles of species and temperature [46]. It should be further noted that atmospheric operation has been a standard for all aforementioned studies, both experimental and numerical.

Key issue in the design and thermal management of microreactors (catalytic or non-catalytic) is the delineation of the regimes where stable combustion can be sustained. The stability of pure gas-phase combustion of methane/air and propane/air mixtures has been investigated numerically [47, 48] in channels with sub-millimeter gap sizes and walls without radical quenching. Therein, 2-D CFD simulations with a one-step gaseous reaction were used to elucidate the impact of heat transfer mechanisms on the energy management at small scales. In pure catalytic combustion over Pt, stability issues were investigated in a 1-mm gap methane-fueled channel with 2-D simulations [39] using a detailed heterogeneous reaction mechanism [49], while another numerical study focused on the stability of propane-fueled catalytic microreactors [41] using a one-step reaction rate expression [50]. The pressure in all aforementioned studies was atmospheric. Despite the large surface-to-volume ratios of microreactors, the contribution of gas-phase reactions cannot always be ignored. Typically, microreactors employ low inlet velocities (less than 1 m/s), which result in sufficiently long residence times for the heat-up and ignition of the gas-phase. Validated heterogeneous and homogeneous kinetics are indispensable inputs in the design of catalytic reactors. The importance of suitable hetero-/homogeneous kinetics should not be understated in microreactor studies. For example, the onset of homogeneous ignition requires gas-phase schemes that can accurately capture the ignition delay characteristics in the presence of the catalytic pathway. The flame anchoring position, in turn, impacts the heat transfer mechanisms in the microreactor structure and the associated extinction limits.

Fundamental studies involving laser-based in situ measurements of thermoscalars in an optically accessible, channel-flow catalytic reactor, have led to the validation and improvement of detailed heterogeneous and homogeneous chemical reaction schemes for fuel-lean methane/air combustion over platinum, at pressures of 1 bar $\leq p \leq$ 16 bar [51, 52]. Even though methane has been extensively studied both experimentally and numerically in catalytic microreactors, propane on the other hand is a fuel of particular interest for micro-energy conversion systems since it liquefies at room temperature and moderate pressures and is commercially available in compact containers for numerous consumer applications. The study of its complete oxidation over catalytic surfaces constitutes a natural first step towards understanding the similar behavior of higher hydrocarbons. Some of the physical characteristics of propane are common among higher hydrocarbons, such as the larger-than-unity Lewis number (in fuel-lean mixtures with air, the Lewis number of propane is $Le \approx 1.8$), a property directly impacting the catalyst surface temperatures [20, 53] and the catalytic fuel conversion rate, and the negative temperature coefficient of the homogeneous ignition delays times under certain operating conditions.

On a fundamental level, the oxidation of propane over noble metals at atmospheric pressure has been studied along with higher and lower hydrocarbons [50, 54], with the former study employing a non-activated dissociative chemisorption step for propane and a temperature-dependent adsorption/desorption reaction set for oxygen. It has been shown that the catalytic reactivity of linear alkanes increases with increasing carbon chain length, due to weaker C–C bond strengths of higher hydrocarbons; in addition, platinum is the most active noble metal for the oxidation of all alkanes except methane [21]. Detailed heterogeneous reaction mechanisms for fuel-lean combustion of higher hydrocarbons have not yet progressed to the same extent as those of methane [55, 56]. Recent studies of fuel-lean propane oxidation over Pt-based catalysts suggested an overall reaction that is first order with respect to propane and zero-th order with respect to oxygen [57]. Accompanying experiments of propane catalytic combustion have mainly been conducted at atmospheric pressure [18, 35, 58].

Until recently, most of the experimental and numerical studies have focused on the steady-state performance of catalytic microburners and microreformers [41, 59–61]. A limited number of studies investigated the transient response and particularly the ignition process (light-off). On the experimental level, the start-up of a propane-fueled, platinum-coated microreactor, with added hydrogen to facilitate light-off, has been demonstrated in [35], while the self-ignition of methanol/air mixtures in a similar reactor coupled to a thermoelectrical device has been studied in [12]. The former work mainly focused on developing a start-up strategy for the specific microreactor, while the latter demonstrated the feasibility of integrating a self-igniting catalytic microcombustor to a power generating system. Similarly, catalytic ignition and extinction in partial oxidation of methane/air mixtures, diluted with exhaust gas recycle, have been investigated experimentally and numerically in rhodium-coated honeycomb reactors for gas-turbine applications [62]. Although the last study did not pertain to microreactors, it nevertheless addressed crucial catalytic ignition requirements in accordance to stringent reactor inlet temperature and residence time limitations. When it comes to modeling of individual catalytic channels, a 1-D transient code was recently employed to study various ignition setups for a Pt-coated microreactor fed with lean propane/air mixtures [63, 64]. Moreover, the majority of transient microreactor simulations emphasized on atmospheric pressure operation, even though moderate pressures up to 5 bar are also of interest. Finally, thermal radiation heat transfer from the hot catalytic walls has not yet been included in transient microreactor models, despite its well-known impact on the light-off of conventional catalytic reactors [65].

1.3 Objectives: Outline

Despite the increased interest for catalytic combustion research during the past few years, many issues remain to be investigated regarding the underlying physico-chemical processes under operational conditions pertinent to small-scale power generation devices. Such issues include the impact high-pressure operation (as proposed in [14–16]) can have on the performance of catalytic microreactors, the implications of initiating gas-phase reactions in the microreactor channels, as well as the proper modeling of all in-channel heat transfer processes, including surface radiation. Moreover, validated hetero-/homogeneous combustion models for realistic fuels (concerning portable applications) such as propane need to be provided, for conditions of pressure-equivalence ratio-reactor temperatures pertinent to microscale power generation. In this work, a number of experimental and numerical investigations were undertaken which facilitated the fundamental and applied understanding of catalytic microreactors operating under micro-gas-turbine conditions. The following are the major objectives set for this work.

- Investigate the catalytic and gas-phase combustion of lean propane/air mixtures on platinum and provide hetero-/homogeneous reaction models validated under conditions of pressure, equivalence ratio and surface temperature relevant to small-scale, gas-turbine-based power generation.
- Design and construct a propane-fueled, catalytic honeycomb combustor to be used in an upscale prototype of a micro-gas-turbine power generation unit, and experimentally assess the thermal power output achieved.
- Use detailed numerical models and chemical schemes to investigate the combustion stability and hetero-/homogeneous chemistry coupling in methane- and propane-fueled catalytic, channel-flow microreactors while accounting for all relevant heat transfer mechanisms, for both transient and steady-state reactor conditions.

In this manuscript, the findings and conclusions of the work initiated by the above stated objectives are presented.

In this chapter, the field of small-scale, portable power generation is introduced. A literature review of the progress in catalytic and gas-phase microreactor research is also presented, along with the objectives of this work.

Chapter 2 introduces the experimental methodology of *in situ* Raman spectroscopy of major gas-phase combustion species and laser induced fluorescence (LIF) of the OH radical, which were employed in an optically accessible, channel-flow, catalytic reactor to study the hetero-/homogeneous combustion of propane on platinum under conditions pertinent to micro-gas-turbines. Moreover, a high-pressure test rig used in evaluating the thermal power output of a mesoscale catalytic honeycomb burner is presented.

The numerical models and computational tools used in the study of a mesoscale catalytic honeycomb combustor and of catalytic microreactors, as well as the associated governing equations, are presented in Chap. 3.

Chapter 4 provides experimental and numerical results obtained from a study on the pure heterogeneous and the coupled hetero-/homogeneous combustion of fuel-lean propane/air mixtures over platinum at pressures 1 bar $\leq p \leq$ 7 bar, fuel-to-air equivalence ratios $0.26 \leq \varphi \leq 0.43$, and catalytic wall temperatures 723 K $\leq T_W \leq$ 1,320 K. Experiments were performed in the optically accessible catalytic channel-flow reactor described in Chap. 2 and involved 1-D Raman measurements of major gas-phase species concentrations across the reactor boundary layer for the assessment of catalytic fuel conversion and planar laser induced fluorescence (LIF) of the OH radical for the determination of homogeneous ignition. The catalytic fuel conversion rate was subsequently assessed as a function of pressure, equivalence ratio and surface temperature [66]. Numerical predictions were carried out with a 2-D elliptic CFD code (introduced in Chap. 3) that included a one-step catalytic reaction for the total oxidation of propane on Pt, an elementary C_3 gas-phase chemical reaction mechanism, and detailed transport. A global catalytic reaction step valid over the entire pressure–temperature—equivalence ratio parameter range has been established, which revealed a $\sim p^{+0.75}$ dependence of the catalytic reactivity on pressure. The aforementioned global catalytic step was further coupled to the elementary gas-phase reaction mechanism in order to simulate homogeneous ignition characteristics in the channel-flow reactor. The predictions reproduced within 10% the measured homogeneous ignition distances at pressures $p \leq 5$ bar, while at $p = 7$ bar the simulations overpredicted the measurements by 19%. The overall model performance suggested that the employed hetero-/homogeneous schemes are suitable for the design of propane-fueled catalytic microreactors.

In the follow-up Chap. 5, the design methodology of a mesoscale catalytic honeycomb burner is presented. A combined experimental and numerical investigation was undertaken for assessing the performance characteristics of a mesoscale, propane-fueled, catalytic combustor, to be used in an integrated, gas-turbine-based, mesoscale (ca. 1,000 W_e) power generation system. Propane was the fuel of choice due to its practicality for portable applications. Parametric numerical studies were carried out in a single catalytic channel, using a 2-D elliptic code with the validated chemical kinetics scheme of Chap. 4 and all relevant heat transfer mechanisms. The predictions identified favorable materials and operating conditions for the catalytic burner under consideration. A subscale model was finally constructed, based on the findings of the parametric study, and subsequently tested in the high-pressure test rig regarding its thermal power output under operational conditions pertinent to small-scale gas turbines. However, heat losses to the environment were responsible for measuring reduced combustor efficiencies at certain operating conditions. A continuum model for the entire monolithic structure complemented the experiments and provided the 2-D temperature field of the burner. Computed exhaust gas temperatures of the monolith were in good agreement with the experiments; findings with the latter tool will allow for future improvements on the combustor design [67].

Chapter 6 presents a numerical study wherein the hetero-/homogeneous steady combustion and the stability limits of methane-fueled catalytic microreactors (Pt-coated) have been investigated in a 1-mm-gap channel at pressures of 1 and 5 bar. Numerical simulations were carried out with a full-elliptic two-dimensional model for the gas and solid phases that included the detailed hetero-/homogeneous reaction schemes provided in [51, 52], heat conduction in the solid wall, surface radiation heat transfer, and external heat losses. Gas-phase chemistry extended the low-velocity stability limits due to the establishment of strong flames and to an even greater degree the high-velocity blowout limits due to the heat release originating primarily from the incomplete homogeneous oxidation of methane. When considering the same mass throughput, the stable combustion envelope at 5 bar was substantially wider than its 1 bar counterpart due to the increased reactivity of both catalytic and gaseous pathways at elevated pressures. Stable combustion could be sustained with solid thermal conductivities at least as low as 0.1 W/mK, while the stability limits reached their larger extent between 20 and 50 W/mK, a range that covers many practical metallic compounds. The stability limits of catalytic microreactors were wider than those reported for non-catalytic systems. Moreover, radiation exchange between channel wall elements was shown to play a dual role, acting from the one side as a net heat loss mechanism by maintaining low reactor temperatures well-within the stable combustion regime (via heat loss to the colder inlet section) and from the other side as a beneficial stabilizing mechanism near the blowout limits by redistributing energy inside the channel [68]. Finally, investigation of smaller confinements showed that gas-phase combustion could be sustained in catalytic microreactors with gaps as low as 0.3 mm.

In Chap. 7, the investigation on combustion stability is extended to propane-fueled catalytic microreactors, using the catalytic and gas-phase chemical reaction schemes of propane combustion on platinum proposed and validated in Chap. 4. The steady hetero-/homogeneous combustion of lean propane/air and methane/air mixtures in a platinum-coated, catalytic plane channel-flow microreactor were investigated at pressures of 1 and 5 bar, channel heights of 1.0 and 0.3 mm, and wall thermal conductivities of 2 and 16 W/mK. Stability limits were assessed as a function of fuel type, inlet velocity, and imposed external heat losses. Parametric studies were performed with a full-elliptic, two-dimensional numerical model employing detailed gas-phase (homogeneous) reaction schemes for both fuels, a detailed heterogeneous (catalytic) reaction scheme for methane and the proposed global reaction step of Chap. 4 for the oxidation of propane on Pt. Comparisons between the stable combustion regimes of methane and propane revealed a strong impact of the fuel molecular transport properties on stability and maximum allowable mass throughput [69]. The higher diffusive transport of methane was critical in maintaining wider high inlet velocity stability limits (blowout) compared to that of propane, despite the higher catalytic and gas-phase reactivity of the latter. On the other hand, at the low velocity limits (extinction), propane exhibited a wider stability envelope. Gas-phase chemistry had a strong impact on the blowout limits, even at channel heights as low as 0.3 mm. For the same mass throughput, smaller channel heights allowed for higher heat losses at the extinction branch of the combustion stability envelope thanks to increased transverse fuel transport while at the same time they yielded narrower blowout limits due to insufficient residence times at higher inlet velocities. The stable combustion regime of propane increased substantially at 5 bar compared to the same mass throughput at 1 bar, owing to a positive $p^{+0.75}$ dependence of the propane catalytic reactivity on pressure. Finally, the role of high wall thermal conductivity in widening the blowout limits for both fuels is demonstrated.

The start-up of methane-fueled catalytic microreactors was studied numerically under the quasisteady assumption for the gaseous phase; results are presented in Chap. 8. Transient simulations have been performed in a plane-channel, methane-fueled microreactor made of either cordierite or FeCr alloy walls and coated with a platinum catalyst. A two-dimensional model for the flow domain was used, which included detailed catalytic and gas-phase chemical reaction mechanisms. In the solid wall, axial heat conduction and surface radiation heat transfer were accounted for. Simulations were performed by varying the inlet pressure, the solid wall thermal conductivity and heat capacity, the inlet velocity, and the equivalence ratio at fuel-lean stoichiometries [70]. The effect of solid material properties as well as the impact of surface radiation and gas-phase chemistry on the ignition (t_{ig}) and steady-state (t_{st}) microreactor times has been assessed. An increase in inlet pressure from 1 to 5 bar induced a $\sim$50% reduction in both t_{ig} and t_{st} owing to the enhancement of the catalytic reactivity with rising pressure. A similar behavior was also attested when increasing the equivalence ratio from 0.4 to 0.6. Reactors with low wall thermal conductivity (cordierite material) exhibited shorter ignition times compared to higher thermal conductivity ones (FeCr alloy) due to the creation of spatially localized hot spots that promoted catalytic ignition. At the same time, the ceramic material required shorter times to reach steady-state. Higher inlet velocities reduced the time required for steady-state, however, at the cost of increased cumulative reactor emissions. Surface radiation heat transfer played a key dual role in the start-up process of low thermal conductivity channels. Radiation increased t_{ig} by removing heat away from the initial hot spot, but from the other side it decreased t_{st} due to a very efficient transfer of heat from the rear to the front of the reactor. Gas-phase chemistry elongated the steady-state times for both ceramic and metallic materials and impacted the emissions of catalytic microreactors.

Finally, in Chap. 9 general conclusions from this work are drawn, along with an outlook of further investigations possible on the general field of catalyzed hydrocarbon oxidation in microreactors.

References

1. Dunn-Rankin D, Martins Leal E, Walther DC (2005) Personal power systems. Prog Energy Combust Sci 31:422–465
2. Fernandez-Pello AC (2002) Micropower generation using combustion: issues and approaches. Proc Combust Inst 29:883–899
3. Peterson RB (2003) In: Faghri M, Sunden (eds) Microscale and nanoscale structures, heat and fluid flow. WIT Press, London
4. Varadan VK (2003) Nanotechnology, MEMS and NEMS and their applications to smart systems and devices. Proc SPIE 5062:20–43
5. Reyntjens S, Puers R (2001) A review of focused ion beam applications in microsystem technology. J Micromech Microeng 11:287–300
6. Pisano A (1997) Microelectromechanical systems (MEMS) program. DARPA/ETO BAA 97–43
7. Tang W (2001) Micro power generation (MPG). DARPA MTP program BAA 01-09
8. Alper J (2002) The battery: not yet a terminal case. Science 296:1224–1226
9. Epstein AH (2004) Millimeter-scale, micro-electro-mechanical systems gas turbine engines. J Eng Gas Turb Power Trans ASME 126:205–226
10. Gomez A, Berry JJ, Roychoudhury S, Coriton B, Huth J (2007) From jet fuel to electric power using a mesoscale, efficient stirling cycle. Proc Combust Inst 31:3251–3259
11. Fu K, Knobloch A, Martinez F, Walther D, Fernandez-Pello AC, Pisano A, Liepmann D, Miyaska K, Maruta K (2001) Design and experimental results of small-scale rotary engines. Proceedings of the IMECE-ASME, New York, 11–16 Nov 2001
12. Karim AM, Federici JA, Vlachos DG (2008) Portable power production from methanol in an integrated thermoelectric/microreactor system. J Power Sources 179:113–120
13. Epstein AH, Senturia SD et al (1997) Micro-heat engines, gas turbines and rocket engines—the MIT microengine project. Paper 97–1773, Twenty-eighth AIAA fluid dynamics conference
14. Isomura K, Murayama M, Teramoto S, Hikichi K, Endo Y, Togo S, Tanaks S (2006) Experimental verification of the feasibility of a 100 W class micro-scale gas turbine at an impeller diameter of 10 mm. J Micromech Microeng 16:S254–S261
15. Peirs J, Waumans T, Vleugels P, Al-Bender F, Stevens T, Verstraete T, Stevens S, D'hulst R, Verstraete D, Fiorini P, Van Der Braembussche R, Driesen J, Puers R, Hendrick P, Baelmans M, Reynaerts D (2007) Micropower generation with microgasturbines: a challenge. J Mech Eng Sci 221:489–500
16. Schneider B, Karagiannidis S, Bruderer M, Dyntar D, Zwyssig C, Guangchun Q, Diener M, Boulouchos K, Abhari RS, Guzzella L, Kolar JW (2005) Ultra-high-energy-density converter for portable power, Power-MEMS 2005, Tokyo, Japan, 28–30 Nov 2005
17. Miesse C, Masel RI, Short M, Shannon M (2005) Experimental observations of methane-oxygen diffusion flame structure in a sub-millimeter microburner. Combust Theory Model 9:77–92
18. Ahn JM, Eastwood C, Sitzki L, Ronney PD (2005) Gas-phase and catalytic combustion in heat recirculating burners. Proc Combust Inst 30:2463–2472
19. Kim KT, Dae HLB, Kwon S (2006) Effects of thermal and chemical surface-flame interaction on flame quenching. Combust Flame 146:19–28
20. Pfefferle WC, Pfefferle LD (1986) Catalytically stabilized combustion. Prog Energy Combust Sci 12:25–41

21. Aryafar M, Zaera F (1997) Kinetic study of the catalytic oxidation of alkanes over nickel, palladium, and platinum foils. Catal Lett 48:173–183
22. Carroni R, Schmidt V, Griffin T (2002) Catalytic combustion for power generation. Catal Today 75:287–295
23. Schneider A, Mantzaras J, Bombach R, Schenker S, Tylli N, Jansohn P (2007) Laser induced fluorescence of formaldehyde and Raman measurements of major species during partial catalytic oxidation of methane with large H_2O and CO_2 dilution at pressures up to 10 bar. Proc Combust Inst 31:1973–1981
24. Smith LL, Karim H, Castaldi MJ, Etemad S, Pfefferle WC (2006) Rich-catalytic lean-burn combustion for fuel-flexible operation with ultra-low emissions. Catal Today 117:438–446
25. Forzatti P (2003) Status and perspectives of catalytic combustion in gas turbines. Catal Today 83:3–18
26. Kim NI, Aizumi S, Yokomori T, Kato S, Fujimori T, Maruta K (2007) Development and scale effects of small Swiss-roll combustors. Proc Combust Inst 31:3243–3250
27. Miesse CM, Masel RI, Jensen CD, Shannon MA, Short M (2004) Submillimeter-scale combustion. AlChE J 50:3206–3214
28. Jin JK, Kwon SJ (2009) Microcatalytic combustion of H_2 on Pt/Al_2O_3-coated nickel foam. Combust Sci Technol 181:211–225
29. Federici JA, Wetzel ED, Geil BR, Vlachos DG (2009) Single channel and heat recirculation microburners: an experimental and computational fluid dynamics study. Proc Combust Inst 32:3011–3018
30. Sitzki L, Borer K, Schuster E, Ronney PD, Wussow S (2001) Combustion in microscale heat-recirculating burners, third Asia-pacific conference on combustion, Seoul, Korea, 24–27 June 2001
31. Maruta K, Takeda K, Sitzki L, Borer K, Ronney PD, Wussow S, Deutschmann O (2001) Catalytic combustion for MEMS power generation, third Asia-Pacific conference on combustion, Seoul, Korea, 24–27 June 2001
32. Vican J, Gajdeczko BF, Dryer FL, Miliusa DL, Aksay IA, Yetter RA (2002) Development of a microreactor as a thermal source for microelectromechanical systems power generation. Proc Combust Inst 29:909–916
33. Choi W, Kwon S, Shin HD (2008) Combustion characteristics of hydrogen-air premixed gas in a sub-millimeter scale catalytic combustor. Int J Hydrogen Energy 33:2400–2408
34. Ibashi W, Groppi G, Forzatti P (2003) Kinetic measurements of CH_4 combustion over a 10% PdO/ZrO_2 catalyst using an annular flow microreactor. Catal Today 83:115–129
35. Norton DG, Vlachos DG (2005) Hydrogen assisted self-ignition of propane/air mixtures in catalytic microburners. Proc Combust Inst 30:2473–2480
36. Kyritsis D, Guerrero-Arias I, Roychoudhury S, Gomez A (2002) Mesoscale power generation by a catalytic combustor using electrosprayed liquid hydrocarbons. Proc Combust Inst 29:965–972
37. Volchko SJ, Sung CJ, Huang YSSJ (2006) Catalytic combustion of rich methane/oxygen mixtures for micropropulsion applications. J Propulsion Power 22:684–693
38. Cheekatamarla PK, Finnerty CM, Robinson CR, Andrews SM, Brodie JA, Lu Y, Dewald PG (2009) Design, integration and demonstration of a 50 W JP8/kerosene fueled portable SOFC power generator. J Power Sources 193:797–803
39. Maruta K, Takeda K, Ahn J, Borer K, Sitzki L, Ronney PD, Deutschmann O (2002) Extinction limits of catalytic combustion in microchannels. Proc Combust Inst 29:957–963
40. Pizza G, Frouzakis CE, Mantzaras J, Tomboulides AG, Boulouchos K (2008) Dynamics of premixed hydrogen/air flames in microchannels. Combust Flame 152:433–450
41. Kaisare NS, Deshmukh SR, Vlachos DG (2008) Stability and performance of catalytic microreactors: simulations of propane catalytic combustion on Pt. Chem Eng Sci 63:1098–1116
42. Ronney PD (2003) Analysis of non-adiabatic heat-recirculating combustors. Combust Flame 135:421–439

43. Stefanidis GD, Vlachos DG (2009) Controlling homogeneous chemistry in homogeneous–heterogeneous reactors: application to propane combustion. Ind Eng Chem Res 48. doi:10.1021/ie801480m
44. Deshmukh SR, Mhadeshwar AB, Vlachos DG (2004) Microreactor modeling for hydrogen production from ammonia decomposition on ruthenium. Ind Eng Chem Res 43:2986–2999
45. Groppi G, Belloli A, Tronconi E, Forzatti P (1995) A comparison of lumped and distributed models of monolith catalytic combustors. Chem Eng Sci 50:2705–2715
46. Mantzaras J, Benz P (1999) An asymptotic and numerical investigation of homogeneous ignition in catalytically stabilized channel flow combustion. Combust Flame 119:455–472
47. Norton DG, Vlachos DG (2003) Combustion characteristics and flame stability at the microscale: a CFD study of premixed methane/air mixtures. Chem Eng Sci 58:4871–4882
48. Norton DG, Vlachos DG (2004) A CFD study of propane/air microflame stability. Combust Flame 138:97–107
49. Deutschmann O, Schmidt R, Behrendt F, Warnatz J (1996) Numerical modeling of catalytic ignition. Proc Combust Inst 26:1747–1754
50. Deshmukh SR, Vlachos DG (2007) A reduced mechanism for methane and one-step rate expressions for fuel-lean catalytic combustion of small alkanes on noble metals. Combust Flame 149:366–383
51. Reinke M, Mantzaras J, Schaeren R, Bombach R, Inauen A, Schenker S (2004) High-pressure catalytic combustion of methane over platinum: in situ experiments and detailed numerical predictions. Combust Flame 136:217–240
52. Reinke M, Mantzaras J, Bombach R, Schenker S, Inauen A (2005) Gas-phase chemistry in catalytic combustion of methane/air mixtures over platinum at pressures of 1–6 bar. Combust Flame 141:448–468
53. Mantzaras J (2006) In: Jiang SZ (ed) Focus on combustion research. Interplay of transport and hetero-/homogeneous chemistry, Nova Publishers, New York, p 241
54. Veser G, Ziauddin M, Schmidt LD (1999) Ignition in alkane oxidation on noble-metal catalysts. Catal Today 47:219–228
55. Deutschmann O, Maier LI, Riedel U, Stroemman AH, Dibble RW (2000) Hydrogen assisted catalytic combustion of methane on platinum. Catal Today 59:141–150
56. Aghalayam P, Park YK, Fernandes N, Papavassiliou V, Mhadeshwar AB, Vlachos DG (2003) A C1 mechanism for methane oxidation on platinum. J Catal 213:23–38
57. Garetto TF, Rincon E, Apesteguia CR (2004) Deep oxidation of propane on Pt-supported catalysts: drastic turnover rate enhancement using zeolite supports. Appl Catal B Environ 48:167–174
58. Bruno C, Walsh PM, Santavicca DA, Sinha N, Yaw Y, Bracco FV (1983) Catalytic combustion of propane/air mixtures on platinum. Combust Sci Technol 31:43–74
59. Stutz MJ, Poulikakos D (2008) Optimum washcoat thickness of a monolith reactor for syngas production by partial oxidation of methane. Chem Eng Sci 63:1761–1770
60. Chao Y-C, Chen G-B, Hsu C-J, Leu T-S, Wu C-Y (2004) Operational characteristics of catalytic combustion in a platinum microtube. Combust Sci Technol 176:1755–1777
61. Kim T, Kwon S (2006) Design, fabrication and testing of a catalytic microreactor for hydrogen production. J Micromech Microeng 16:1760–1768
62. Schneider A, Mantzaras J, Eriksson S (2008) Ignition and extinction in catalytic partial oxidation of methane–oxygen mixtures with large H_2O and CO_2 dilution. Combust Sci Technol 180:89–126
63. Stefanidis GD, Kaisare NS, Vlachos DG (2008) Modeling ignition in catalytic microreactors. Chem Eng Technol 31:1170–1175
64. Kaisare NS, Stefanidis GD, Vlachos DG (2009) Comparison of ignition strategies for catalytic microburners. Proc Combust Inst 32:3027–3034
65. Boehman AL (1998) Radiation heat transfer in catalytic monoliths. AIChE 44:2745–2755
66. Karagiannidis S, Mantzaras J, Bombach R, Schenker S, Boulouchos K (2009) Experimental and numerical investigation of the hetero-/homogeneous combustion of lean propane/air mixtures over platinum. Proc Combust Inst 32:1947–1955

67. Karagiannidis S, Marketos K, Mantzaras J, Schaeren R, Boulouchos K (2010) Experimental and numerical investigation of a propane-fueled, catalytic, mesoscale combustor. Catal Today 155:108–115
68. Karagiannidis S, Mantzaras J, Jackson G, Boulouchos K (2007) Hetero-/homogeneous combustion and stability maps in methane-fueled catalytic microreactors. Proc Combust Inst 31:3309–3317
69. Karagiannidis S, Mantzaras J, Boulouchos K (2010) Stability of hetero-/homogeneous combustion in propane- and methane-fueled catalytic 2011 microreactors: channel confinement and molecular transport effects. Proc Combust Inst 33:3241–3249.
70. Karagiannidis S, Mantzaras J (2010) Numerical investigation on the start-up of methane-fueled catalytic microreactors. Combust Flame 157:1400–1413

Chapter 2
Experimental Setup

2.1 High-Pressure, Optically Accessible, Catalytic Channel-Flow Reactor

The test rig employed in this study consisted of a high-pressure cylindrical steel vessel with a length of 1.8 m and an internal diameter of 0.28 m. Visual inspection and accessibility of the reactor assembly was achieved via a 50 mm diameter quartz window at the rear flange of the vessel and two 350 mm long and 50 mm high quartz windows at the vessel sides. The test setup (Fig. 2.1) consisted of a channel-flow catalytic reactor, which was mounted inside the high-pressure cylindrical vessel. The reactor comprised two horizontal Si[SiC] plates with a length (x) of 300 mm, width (z) of 104 mm and thickness of 9 mm; the plates were positioned 7 mm apart (y). The other two sides of the reactor were formed by two 3-mm-thick vertical quartz windows.

The inner surfaces of the Si[SiC] plates were coated via plasma vapor deposition with a 1.5 μm thick Al_2O_3 non-porous layer, followed by a 2.2 μm thick platinum layer. BET and CO-chemisorption measurements verified the absence of porous structures in the catalyst layer [1]. The surface temperatures along the x–y symmetry plane were monitored with S-type thermocouples (12 for each plate), which were embedded 0.9 mm beneath the catalyst layer. In order to facilitate the ensuing catalytic reactivity studies, a kinetically-controlled fuel conversion away from the mass-transport-limit was necessary. To avoid transport limitations originating from the high catalytic reactivity of propane over platinum, a coupled cooling/heating arrangement was adopted to control the surface temperatures: in a fashion similar to earlier hydrogen catalytic combustion studies [2], the reactor entry section was water-cooled to maintain a low catalytic reactivity while the rear of the reactor ($x > 100$ mm) was heated by two resistive heaters placed above the ceramic plates.

S. Karagiannidis, *Catalytic Microreactors for Portable Power Generation*,
Springer Theses, DOI: 10.1007/978-3-642-17668-5_2,

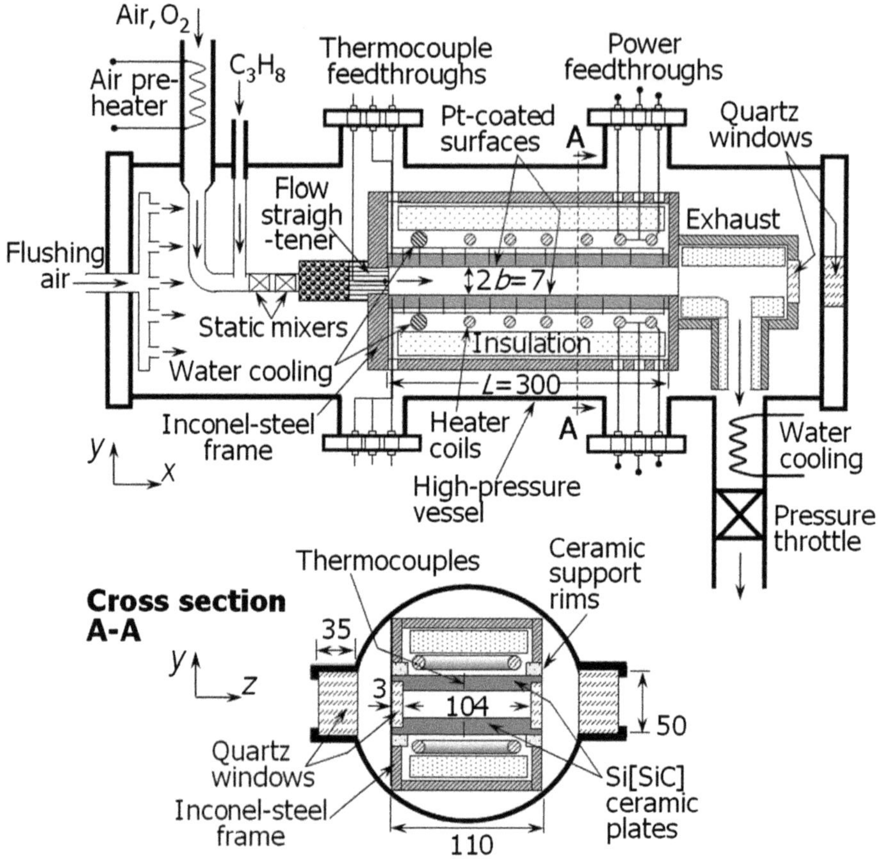

Fig. 2.1 Schematic of the high-pressure steel vessel, mounted with the optically accessible, channel-flow catalytic reactor. All distances are in mm

A compressor provided dry air, which was preheated and mixed with propane (Grade 3.5 purity) in two sequential static mixers. The preheated propane/air mixture was driven into the reactor through a 50-mm long inert rectangular honeycomb section that provided a uniform inlet velocity profile. The reactor inlet temperature was monitored with a thermocouple placed downstream of the honeycomb section. The high-pressure vessel was fitted with two 350-mm long and 35-mm thick quartz windows (see Fig. 2.2) which maintained optical accessibility from both reactor sides. Two additional quartz windows located at the exhaust section of the vessel and the reactor outlet provided a counterflow optical access for the LIF experiments. Apart from propane/air, experiments with propane/air/oxygen mixtures have also been carried out.

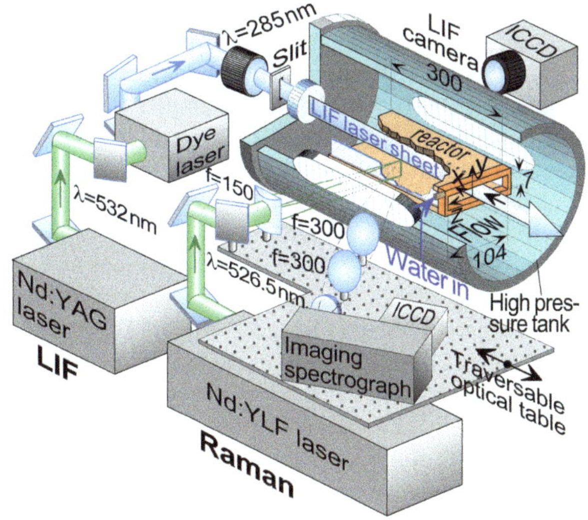

Fig. 2.2 Test rig and Raman/LIF setup

2.2 Laser Diagnostics: Raman Spectroscopy and OH-LIF

The LIF/Raman set-up is shown in Fig. 2.2. The particularly low volumetric content of fuel in lean propane/air combustion necessitated the use of a dedicated high power laser for the Raman measurements. The 526.5 nm radiation of a high repetition rate, frequency-doubled Nd:YLF pulsed laser (Quantronix Darwin Duo), operated at 1.5 to 2 kHz, with a pulse duration and energy of 130 ns and 37 to 43 mJ, respectively, provided the light source for the Raman measurements. Given the laminar and steady operating conditions, the signal of 20,000–40,000 pulses was integrated on the detector chip when acquiring an image. Six of these images were subsequently averaged, such that up to 9 kJ of laser light was used for a single Raman spectrum. The signal-to-noise ratio increased by a factor of 20 compared to the previous methane/air Raman arrangement [3, 4]; moreover, the danger of dielectric gas breakdown was eliminated due to the resulting lower intensities at the focal line. The 526.5 nm beam was focused through the vessel and reactor side-windows into a vertical line ($\sim$0.3 mm thick) by an $f = 150$ mm cylindrical lens. The focal line spanned the 7 mm channel separation and was offset laterally ($z = 15$ mm) to increase the collection angle and minimize thermal beam steering [4]. Two $f = 300$ mm lenses collected the scattered light at a 50° angle with respect to the incident optical path and focused it to the entrance slit of a 25 cm imaging spectrograph (Chromex-250i) equipped with an intensified CCD camera (Princeton Instruments PI-MAX1024GIII). The 1024- and 256-pixel-long CCD dimensions corresponded to wavelength and transverse distance, respectively. The effective Raman cross sections, which included transmission efficiencies, were evaluated by recording the signals of pure propane, air, and completely burnt gases of known composition. Raman data for the major combustion species C_3H_8, H_2O, N_2, O_2 and CO_2 were acquired at different positions by traversing axially a table supporting the sending and collecting optics and also the Nd:YLF laser (Fig. 2.2). The 250-pixel-long 7 mm channel height was binned to 63 pixels.

For the OH-LIF, the 532 nm radiation of a frequency-doubled Nd:YAG laser (Quantel TDL90 NBP2UVT3) pumped a tunable dye laser (Quantel TDL90); its frequency-doubled radiation (285 nm) had a pulse energy of 0.5 mJ, low enough to avoid saturation of the $A(v = 1) \leftarrow X(v' = 0)$ transition. The 285 nm beam was transformed into a laser sheet by a cylindrical lens telescope and a 1 mm slit mask, which propagated counterflow, along the x–y symmetry plane (Fig. 2.2). The fluorescence of both (1-1) and (0-0) transitions at 308 and 314 nm, respectively, was collected at 90° (through the reactor and vessel side-windows) with an intensified CCD camera (LaVision Imager Compact HiRes IRO, 1,392 × 1,024 pixels binned to 696 × 512). A 120 × 7 mm^2 section of the combustor was imaged on a 600 × 34 pixel CCD-area. The camera was traversed axially to map the 300 mm reactor extent; at each measuring location 400 images were averaged. A picture of the actual test rig mounted with optics and Raman set-up is provided in Fig. 2.3, along with pictures of the platinum-coated plates, complete with the embedded thermocouples.

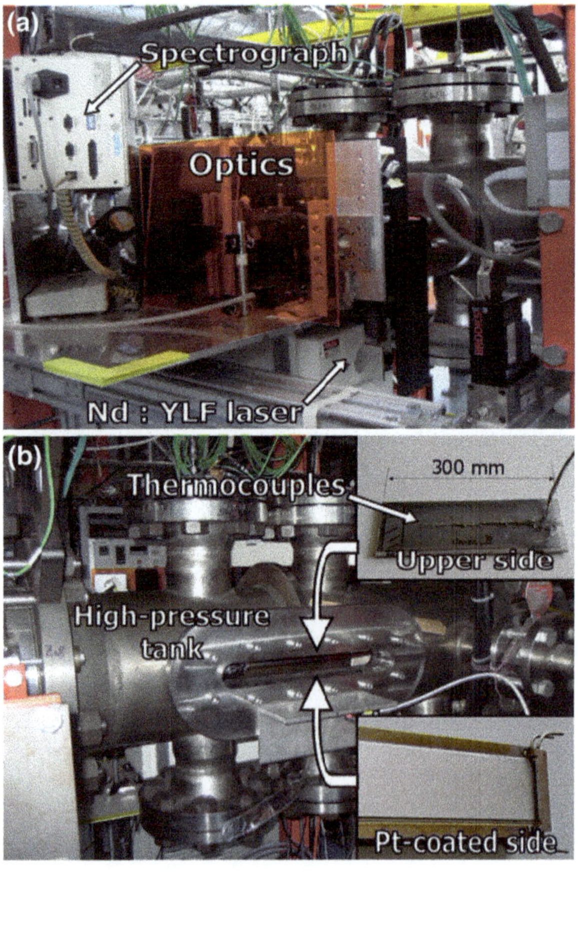

Fig. 2.3 Optics, spectrograph and Nd:YLF laser mounted on the test rig (**a**), and side view of the high-pressure test rig (**b**), along with pictures of the platinum-coated and thermocouple-fitted sides of a reactor plate

2.3 Subscale Catalytic Honeycomb Combustor

The honeycomb catalytic burner tested in this work was a subscale version of the actual catalytic combustor proposed for the micro-gas-turbine-based power unit, and was embedded inside the high-pressure vessel (see Fig. 2.4). It comprised a 35 mm inner-diameter, 75 mm long (L) and 1.5 mm thick steel tube, wherein alternating flat and corrugated FeCr-alloy foils (with thickness $\delta = 50$ μm) were rolled up forming a honeycomb structure with a channel density of 400 cpsi.

Visual inspection of the subscale unit revealed an overall good cross-section uniformity for the catalytic channels (see Fig. 2.4). The unit tested in this study differed from the proposed mesoscale catalytic combustor only in its inner diameter (all other geometric parameters were kept constant), with the former unit having a radial dimension ~42% smaller than the latter. With the number of catalyst-coated channels being proportional to the honeycomb cross sectional area, the power output of the subscale unit was reduced nearly threefold compared to the mesoscale unit; heat losses from the outer combustor surface were accordingly impacted due to the reduced surface area of the subscale combustor.

The cross section of each channel was triangular with rounded corners and the equivalent hydraulic radius was $r_h = 0.507$ mm. The FeCr-alloy foils were coated with a porous 5%-wt Pt/Al_2O_3 washcoat of ~15 μm thickness. Details on the

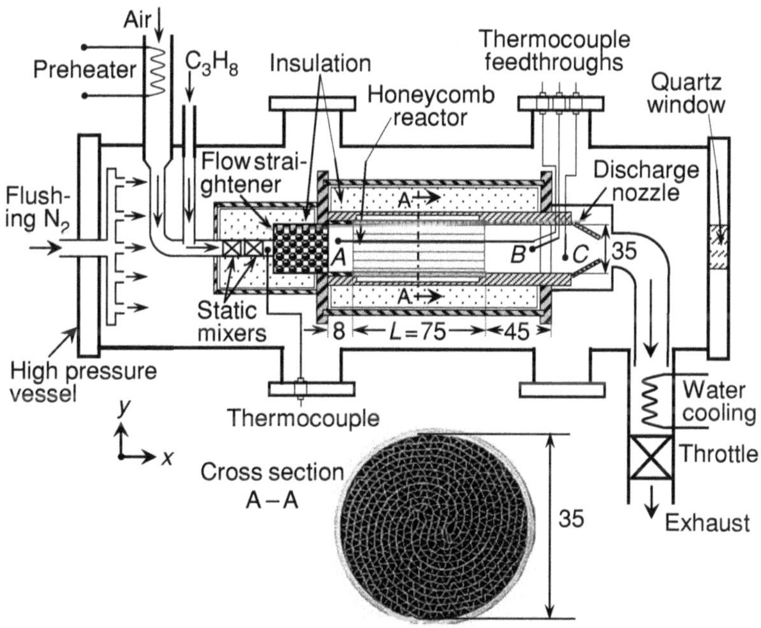

Fig. 2.4 High-pressure test rig fitted with the honeycomb catalytic combustor. Points A, B and C denote the thermocouple positions. Cross-section A–A presents a top-down view of the subscale catalytic combustor. All distances are in mm

catalytic washcoat preparation can be found elsewhere [5]. BET and H_2-chemisorption measurements of the fresh Pt-coated foils determined the active-to-geometrical surface ratio of the catalytic washcoat; this value was found to be equal to $B = 15$ and was later used as input in the computational model developed to simulate the entire honeycomb monolith structure.

High pressure bottles supplied technical grade (99.5% purity) propane, and dry air was supplied by a compressor. Two Brooks mass-flow meters regulated the corresponding flows. The air flow was preheated by a 3 kW resistive heater and then mixed with room temperature propane in two sequential static mixers. A follow-up 40 mm long packing of 2 mm diameter ceramic spheres straightened the flow. The reactor was mounted inside an insulated (using a 30-mm-thick fiber ceramic material) cylindrical steel frame and was affixed 8 mm downstream of the flow straightener (Fig. 2.4). Only the first and last 2 mm of the reactor contacted the steel frame, while in the remaining length a 1-mm-thick annular air-cushion was available. In order to regulate the temperature inside the high pressure vessel, air flowed in the free volume between the reactor and the vessel. The inlet and outlet reactor temperatures were monitored with three 50-μm thick K-type (Ni/Cr–Ni/Al) sheathed thermocouples (designated A through C in Fig. 2.4). The thermocouple beads were positioned at $x = -15$, 95 and 125 mm, with $x = 0$ denoting the beginning of the combustor monolith. The absolute accuracy of the gas temperature measurements was ~10 K for the hot outlet and ~6 K for the inlet. A picture of the honeycomb catalytic combustor tested is provided in Fig. 2.5.

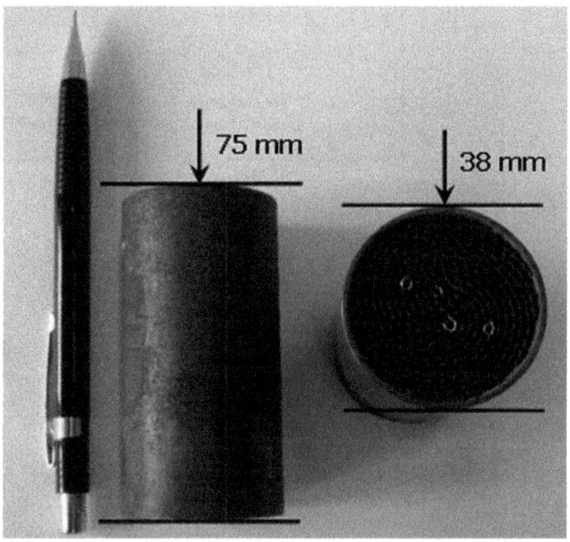

Fig. 2.5 Subscale, catalytic, honeycomb combustor

References

1. Reinke M, Mantzaras J, Bombach R, Schenker S, Inauen A (2005) Gas-phase chemistry in catalytic combustion of methane/air mixtures over platinum at pressures of 1 bar to 16 bar. Combust Flame 141:448–468
2. Appel C, Mantzaras J, Schaeren R, Bombach R, Inauen A, Kaeppeli B, Hemmerling B, Stampanoni A (2002) An experimental and numerical investigation of homogeneous ignition in catalytically stabilized combustion of hydrogen/air mixtures over platinum. Combust Flame 128:340–368
3. Reinke M, Mantzaras J, Schaeren R, Bombach R, Inauen A, Schenker S (2004) High-pressure catalytic combustion of methane over platinum: in situ experiments and detailed numerical predictions. Combust Flame 136:217–240
4. Reinke M, Mantzaras J, Schaeren R, Bombach R, Inauen A, Schenker S (2005) Homogeneous ignition of CH_4/air and H_2O- and CO_2-diluted CH_4/O_2 mixtures over platinum; an experimental and numerical investigation at pressures up to 16 bar. Proc Combust Inst 30:2519–2527
5. Eriksson S, Wolf M, Schneider A, Mantzaras J, Raimondi F, Boutonnet M, Järas S (2006) Fuel-rich catalytic combustion of methane in zero emissions power generation processes. Catal Today 117:447–453

Chapter 3
Numerical Models

3.1 Two-Dimensional, Full-Elliptic Flow Model

A full-elliptic, two-dimensional CFD code [1–4] has been used to simulate the flow domain in a plane channel configuration having length L, height $2b$ and wall thickness δ. The full channel-flow reactor height was modeled in the fundamental studies of lean propane/air combustion on platinum, while in all other cases only half of the channel domain was modeled due to symmetry.

The governing equations in Cartesian coordinates for laminar steady flows are:
Continuity equation:

$$\frac{\partial(\rho u)}{\partial x} + \frac{\partial(\rho v)}{\partial y} = 0. \tag{3.1}$$

Momentum equations:

$$\frac{\partial(\rho uu)}{\partial x} + \frac{\partial(\rho vu)}{\partial y} + \frac{\partial p}{\partial x} - \frac{\partial}{\partial x}\left[2\mu\frac{\partial u}{\partial x} - \frac{2}{3}\mu\left(\frac{\partial u}{\partial x} + \frac{\partial v}{\partial y}\right)\right] - \frac{\partial}{\partial y}\left[\mu\left(\frac{\partial u}{\partial y} + \frac{\partial v}{\partial x}\right)\right] = 0, \tag{3.2}$$

$$\frac{\partial(\rho uv)}{\partial x} + \frac{\partial(\rho vv)}{\partial y} + \frac{\partial p}{\partial y} - \frac{\partial}{\partial x}\left[\mu\left(\frac{\partial v}{\partial x} + \frac{\partial u}{\partial y}\right)\right] - \frac{\partial}{\partial y}\left[2\mu\frac{\partial v}{\partial y} - \frac{2}{3}\mu\left(\frac{\partial u}{\partial x} + \frac{\partial v}{\partial y}\right)\right] = 0. \tag{3.3}$$

Energy equation:

$$\frac{\partial(\rho uh)}{\partial x} + \frac{\partial(\rho vh)}{\partial y} + \frac{\partial}{\partial x}\left(\rho\sum_{k=1}^{Kg} Y_k h_k V_{k,x} - k\frac{\partial T}{\partial x}\right) + \frac{\partial}{\partial y}\left(\rho\sum_{k=1}^{Kg} Y_k h_k V_{k,y} - k\frac{\partial T}{\partial y}\right) = 0. \tag{3.4}$$

S. Karagiannidis, *Catalytic Microreactors for Portable Power Generation*,
Springer Theses, DOI: 10.1007/978-3-642-17668-5_3,

Gas phase species equations:

$$\frac{\partial(\rho uYk)}{\partial x}+\frac{\partial(\rho vYk)}{\partial y}+\frac{\partial}{\partial x}(\rho Y_k V_{k,x})+\frac{\partial}{\partial y}(\rho Y_k V_{k,y})-\dot{\omega}_k W_k=0, \quad k=1,\ldots,K_g. \tag{3.5}$$

Surface species coverage equations:

$$\sigma_m\frac{\dot{s}_m}{\Gamma} = 0, \quad m=1,\ldots,M_s. \tag{3.6}$$

The species diffusion velocities $\vec{V}_k$. were computed using mixture average diffusion, including thermal diffusion for the light species H and H_2 [5]:

$$\vec{V}_k=-D_{km}\nabla[\ln(Y_k\bar{W}/W_k)]+\left[D_k^T W_k/(\rho Y_k\bar{W})\right]\nabla(\ln T) \tag{3.7}$$

Finally, the ideal gas and caloric state laws closed the system of equations:

$$p=\frac{\rho RT}{\bar{W}} \text{ and } h_k=h_k^o(T_o)+\int_{T_o}^{T} c_{p,k}\,dT. \tag{3.8}$$

In Chap. 4, the boundary conditions for the energy equation at the gas-wall interfaces ($y=0$ and $y=2b$) were:

$$T(x,y=0)=T_{W,L}(x) \tag{3.9}$$

$$T(x,y=2b)=T_{W,U}(x), \tag{3.10}$$

with $T_{W,L}(x)$ and $T_{W,U}(x)$ the thermocouple-measured temperature distributions of the lower and upper walls, respectively.

In Chaps. 5–7, a 2-D approach was adopted for the solid substrate (channel wall). Solution was obtained for the steady solid heat conduction equation:

$$k_s\frac{\partial^2 T_W}{\partial x^2}+k_s\frac{\partial^2 T_W}{\partial y^2}=0, \tag{3.11}$$

with the interfacial energy balance between solid and gas-phase ($y=b$) in this case being:

$$\dot{q}_{rad}-k\left.\frac{\partial T_W}{\partial y}\right|_{y=b-}+k_s\left.\frac{\partial T_W}{\partial y}\right|_{y=b+}+\sum_{k=1}^{Kg}(\dot{s}_k h_k W_k)=0. \tag{3.12}$$

In Chap. 8, a one-dimensional approach was used for the channel wall. In this case, the transient 1-D energy balance in the solid becomes:

$$\left(\rho_s c_s\frac{\partial T_W}{\partial t}-k_s\frac{\partial^2 T_W}{\partial x^2}\right)\delta-\left(q_{rad}-k\left.\frac{\partial T}{\partial y}\right|_{y=b}+\sum_{k=1}^{Kg}(\dot{s}_k h_k W_k)\right)=0. \tag{3.13}$$

The net radiation method for diffuse-gray areas [6] accounted for radiation exchange between the discretized channel wall elements themselves, and between each wall element and the inlet and outlet channel enclosures; gas radiative emission and adsorption were not considered, given the small optical paths and large nitrogen content.

For the k-th channel surface element, the radiation balance becomes:

$$\frac{1}{\varepsilon_k} q_k = \sum_{j=1}^{N+2} F_{k-j}\sigma\left(T_k^4 - T_j^4\right) + \sum_{j=1}^{N+2}\left(\frac{1-\varepsilon_j}{\varepsilon_j}\right) F_{k-j} q_j, \tag{3.14}$$

with $q_k = q_{rad}$ for each channel wall element and j running over the N wall elements as well as the inlet ($j = N + 1$) and outlet ($j = N + 2$).

The interfacial boundary conditions for the gas phase species become:

$$(\rho Y_k V_k)_{y=b} + B\dot{s}_k W_k = 0, \quad k = 1, \ldots, K_g, \tag{3.15}$$

with B (active-to-geometrical surface ratio) set to $B = 1$ in all subsequent calculations, except for Chap. 5 were an experimentally determined value of $B = 15$ was used.

In the case of infinitely-wide parallel plates, the configuration factors F_{k-j} between channel wall surface elements are [3]:

$$\begin{aligned} F_{k-j} = \frac{1}{2(x_{k+1} - x_k)} \Bigg(& \left[(2b)^2 + (x_{k+1} - x_j)^2\right]^{1/2} \\ & - \left[(2b)^2 + (x_{k+1} - x_{j+1})^2\right]^{1/2} + \left[(2b)^2 + (x_k - x_{j+1})^2\right]^{1/2} \\ & - \left[(2b)^2 + (x_k - x_j)^2\right]^{1/2} \Bigg) \quad j = 1, \ldots, N, \end{aligned} \tag{3.16}$$

while the configuration factors between each element k and the inlet and outlet enclosures ($j = N + 1$ and $N + 2$ respectively) are [3]:

$$\begin{aligned} F_{k-(N+1)} = \frac{1}{2(x_{k+1} - x_k)} \Bigg(& \left[(2b)^2 + (x_k)^2\right]^{1/2} - \left[(2b)^2 + (x_{k+1})^2\right]^{1/2} \\ & + (x_{k+1} - x_k) \Bigg) \end{aligned} \tag{3.17}$$

$$\begin{aligned} F_{k-(N+2)} = \frac{1}{2(x_{k+1} - x_k)} \Bigg(& \left[(2b)^2 + (x_{k+1} - L)^2\right]^{1/2} - \left[(2b)^2 + (x_k - L)^2\right]^{1/2} \\ & + (x_{k+1} - x_k) \Bigg) \end{aligned} \tag{3.18}$$

where the element k occupies the length between x_k and x_{k+1}.

Radiative boundary conditions were finally applied to the vertical front and rear solid wall faces:

$$k_s \left.\frac{\partial T_W}{\partial x}\right|_{x=0} = \varepsilon_{eff,\mathrm{IN}}\sigma\left[T_W^4(x=0) - T_{\mathrm{IN}}^4\right], \tag{3.19}$$

$$-k_s \left.\frac{\partial T_W}{\partial x}\right|_{x=L} = \varepsilon_{eff,\mathrm{OUT}}\sigma\left[T_W^4(x=L) - T_{\mathrm{OUT}}^4\right], \tag{3.20}$$

where T_{OUT} is the average gas temperature at the outlet, the assumption being that the inlet and outlet manifolds are at thermal equilibrium with the corresponding inlet/outlet gas temperatures. The effective solid emissivity ε_{eff} towards the inlet and outlet channel enclosures is [7]:

$$\varepsilon_{eff,\mathrm{IN}} = \left[\left(\frac{1}{\varepsilon}\right) + \left(\frac{r_{\mathrm{IN}}}{\varepsilon_{\mathrm{IN}}}\right) - r_{\mathrm{IN}}\right]^{-1} \text{ and } \varepsilon_{eff,\mathrm{OUT}} = \left[\left(\frac{1}{\varepsilon}\right) + \left(\frac{r_{\mathrm{OUT}}}{\varepsilon_{\mathrm{OUT}}}\right) - r_{\mathrm{OUT}}\right]^{-1}, \tag{3.21}$$

with r_{IN}, r_{OUT} the ratios of the channel vertical front and rear areas over the areas they subtend at the inlet and outlet, respectively. Since in the subsequent simulations the vertical wall faces had very small heights in all cases considered (δ = 50–100 μm), even a small inlet/outlet enclosure area justified the use of $\varepsilon_{eff,\mathrm{IN}} = \varepsilon_{eff,\mathrm{OUT}} = \varepsilon$.

The coupled set of flow and solid equations (Eqs. (3.1)–(3.8), (3.11)) were solved simultaneously. A finite volume scheme was adopted for the spatial discretization of the flow equations and solution was obtained with a SIMPLER method for the pressure-velocity field [8]. In the case of transient simulation, the 1-D transient solid energy equation was solved with a second order accurate, fully implicit scheme by using a quadratic backward time discretization [9]. The coupled flow and solid phases were solved iteratively and convergence was achieved at each time step when the solid temperature did not vary at any position along the wall by more than 10^{-5} K.

3.2 Combustor Monolith Continuum Model

In the continuum model approach, the two-dimensional temperature equation under the finite volume approach for the entire monolithic structure of the honeycomb catalytic combustor was solved in cylindrical coordinates:

$$\frac{\partial(\rho_c c_c T)}{\partial t} = \frac{\partial^2(k_{c,x}T)}{\partial x^2} + \frac{1}{r}\frac{\partial}{\partial r}\left[r\frac{\partial(k_{c,r}T)}{\partial r}\right] + q, \tag{3.22}$$

with the effective monolith properties of density ρ_c, heat capacity c_c and thermal conductivity $k_{c,x}$, $k_{c,r}$ assuming weighted average values, estimated from solid material and gas-phase properties [10]. Variable thermal conductivity in the axial and radial directions was also accounted for [11]. The coupling to the reactive flow fields of the individual catalytic channels in the monolith was achieved with the use of the heat source term q, calculated for a representative number of channels, which accounted for the heat transfer between gas-phase and solid by the change in the integral (across the channel diameter $2r$) enthalpy flux in the gas-phase:

$$q = -\sigma\left(\frac{\partial \dot{H}_g}{\partial x}\right). \tag{3.23}$$

By choosing a time step for the transient solid temperature equation long enough for gas-phase equilibration, the discretized, time-independent set of the Navier–Stokes equations under the boundary layer approximation can be solved for the flow field inside the catalytic channels [12] (quasisteady assumption for the gas-phase) using the CRESLAF package [13]. The applicability of the boundary layer approach in catalytic combustion at sufficiently large Reynolds numbers (Re $\geq$ 20) has already been demonstrated [14]. The simplified equations thus become:

Continuity equation:

$$\frac{\partial(\rho u)}{\partial x} + \frac{1}{r}\frac{\partial(r\rho v)}{\partial r} = 0 \tag{3.24}$$

Momentum equations:

$$\frac{\partial(\rho uu)}{\partial x} + \frac{\partial(\rho vu)}{\partial r} = -\frac{\partial p}{\partial x} + \frac{1}{r}\frac{\partial}{\partial r}\left(\mu\frac{\partial(ur)}{\partial r}\right) \tag{3.25}$$

$$\frac{\partial p}{\partial r} = 0 \tag{3.26}$$

Gas phase species equations:

$$\frac{\partial(\rho uYk)}{\partial x} + \frac{\partial(\rho vYk)}{\partial r} + \frac{1}{r}\frac{\partial}{\partial r}(\rho Y_k V_{k,y} r) - \dot{\omega}_k W_k = 0, \quad k = 1,\ldots,K_g. \tag{3.27}$$

Energy equation:

$$\frac{\partial(\rho u c_p T)}{\partial x} + \frac{\partial(\rho v c_p T)}{\partial r} = \frac{1}{r}\frac{\partial}{\partial r}\left(k\frac{\partial(rT)}{\partial r}\right) - \frac{1}{r}\frac{\partial}{\partial r}\left(\rho r\sum_{k=1}^{Kg} Y_k V_{k,y} c_{p,k}\frac{\partial T}{\partial r}\right) - \sum_{k=1}^{Kg}\dot{\omega}_k W_k h_k. \tag{3.28}$$

After each time step, the resulting axial temperature profiles from Eq. (3.22) at the radial positions corresponding to the positions of the representative channels simulated were applied as boundary conditions for the individual flow field channel calculations.

References

1. Schneider A, Mantzaras J, Eriksson S (2008) Ignition and extinction in catalytic partial oxidation of methane-oxygen mixtures with large H_2O and CO_2 dilution. Combust Sci Technol 180:89–126
2. Karagiannidis S, Mantzaras J, Jackson G, Boulouchos K (2007) Hetero-/homogeneous combustion and stability maps in methane-fueled catalytic microreactors. Proc Combust Inst 31:3309–3317
3. Dogwiler U, Benz P, Mantzaras J (1999) Two-dimensional modeling for catalytically stabilized combustion of a lean methane-air mixture with elementary homogeneous and heterogeneous chemical reactions. Combust Flame 116:243–258
4. Schneider A, Mantzaras J, Jansohn P (2006) Experimental and numerical investigation of the catalytic partial oxidation of CH_4/O_2 mixtures diluted with H_2O and CO_2 in a short contact time reactor. Chem Eng Sci 61:4634–4646
5. Kee RJ, Dixon-Lewis G, Warnatz J, Coltrin ME, Miller JA (1996) A Fortran computer code package for the evaluation of gas-phase multicomponent transport properties, Report No. SAND86-8246; Sandia National Laboratories
6. Siegel R, Howell JR (1981) Thermal radiation heat transfer. Hemisphere, New York, p 271
7. Incropera FP, Dewitt DP (1996) Introduction to heat transfer. Wiley, New York
8. Patankar SV (1980) Numerical heat transfer and fluid flow. Hemisphere Publ. Corp., New York
9. Ferziger JH, Petric M (1999) Computational methods for Fluid Dynamics. Springer, New York
10. Tischer S, Correa C, Deutschmann O (2001) Transient three-dimensional simulation of a catalytic combustion monolith using detailed models for heterogeneous and homogeneous reactions and transport phenomena. Catal Today 69:57–62
11. Groppi G, Tronconi E (1996) Continuous vs. discrete models on nonadiabatic monolith catalysts. AIChE 42:2382–2387
12. Brenan KE, Campbell SL, Petzold LR (1989) Numerical solution of initial-value problems in differential-algebraic equations. Prentice Hall, New York
13. Coltrin ME, Moffatt HK, Kee RJ, Rupley FM (1996) CRESLAF (Version 4.0)*: a Fortran Program for Modeling Laminar, Chemically Reacting, Boundary-Layer Flow in Cylindrical or Planar Channels, SAND93-0478; Sandia National Laboratories
14. Raja LL, Kee RJ, Deutschmann O, Warnatz J, Schmidt LD (2000) A critical evaluation of Navier-Stokes, boundary-layer, and plug-flow models of the flow and chemistry in a catalytic-combustion monolith. Catal Today 59:47–60

Chapter 4
Experimental and Numerical Investigation of the Hetero-/Homogeneous Combustion of Lean Propane/Air Mixtures Over Platinum

4.1 Preface

In the present chapter a combined experimental and numerical investigation is presented on the catalytic combustion of fuel-lean C_3H_8/air mixtures over platinum at pressures $1 \leq p \leq 7$ bar, arange of interest to microreactors and small-scale industrial turbines. Experiments have been carried out in the optically accessible, channel-flow catalytic reactor at propane-to-air equivalence ratios ranging between 0.26 and 0.43. The heterogeneous reactivity is assessed with Raman spectroscopy of major gas-phase species concentrations across the channel boundary layer, and the onset of homogeneous ignition is monitored with planar laser induced fluorescence (LIF) of the OH radical. Simulations are performed with the full-elliptic 2-D CFD code. The main objectives are to assess the catalytic reactivity of propane and its pressure dependence by constructing an appropriate global reaction step and then to couple this step with a detailed homogeneous reaction mechanism so as to reproduce the measured homogeneous ignition characteristics.

4.2 Experimental Setup and Numerical Model

The test-rig used in this work consisted of a channel-flow catalytic reactor, which was mounted inside a high-pressure cylindrical vessel. A detailed description has been given in Sect. 2.1.

The lack of detailed surface reaction mechanisms for propane on platinum necessitated the use of a single-step catalytic reaction coupled to a detailed gas-phase reaction mechanism. The global reaction step of Garetto et al. [1] has been developed for the total oxidation of propane to H_2O and CO_2 over Pt at

S. Karagiannidis, *Catalytic Microreactors for Portable Power Generation*, Springer Theses, DOI: 10.1007/978-3-642-17668-5_4,

atmospheric pressure, for a range of equivalence ratios encompassing the conditions of this work:

$$\dot{s}_{C_3H_8} = A \times T_W^{1.15} \times \exp(-E_a/RT_W) \times [C_3H_8]_W^a. \quad (4.1)$$

The catalytic rate $\dot{s}_{C_3H_8}$ is in (mol/cm^2 s), $A = 93.2\ K^{-1.15} cm^{1.45} mol^{-0.15}\ s^{-1}$, $E_a = 71.128$ kJ/mole, $a = 1.15$ and the concentration of propane $[C_3H_8]$ is in (mol/cm^3). Finally, the subscript w denotes conditions at the gas-wall interface.

An optimized mechanism for homogeneous combustion of C_1–C_3 species by Qin et al. (70 species, 14 irreversible and 449 reversible reactions) [2] was employed for modeling gas-phase chemistry. Thermodynamic data were included in the provided scheme. Surface and gas-phase reaction rates were evaluated with Surface-CHEMKIN [3] and CHEMKIN [4] respectively. Mixture-average diffusion was the transport model, using the CHEMKIN transport database [5].

The steady, full-elliptic, 2-D laminar CFD code presented in Sect. 3.1 was used in the simulations (for details see also [6]). An orthogonal staggered grid of 450 × 140 points (x and y, respectively) was sufficient to give a grid independent solution for the 300 × 7 mm^2 channel domain. Uniform inlet profiles were applied for temperature, axial velocity and species mass fractions. The interfacial energy boundary conditions were prescribed wall temperature profiles; these profiles were polynomial curves fitted through the thermocouple measurements of the upper and lower walls. No-slip was applied for both velocity components at the walls ($y = 0$ and 7 mm), while zero-Neumann conditions were used at the outlet.

4.3 Results and Discussion

The experimental conditions are provided in Table 4.1.

Table 4.1 Experimental conditions

Case	p	φ	T_{IN}	U_{IN}	Re_{IN}	N_2 vol.%
1	1	0.30	398	1.45	827	55
2	3	0.27	400	1.10	1,831	60
3	5	0.27	399	0.66	1,836	60
4	7	0.26	398	0.55	2,162	60
5	3	0.31	446	1.58	2,171	77
6	5	0.23	448	0.94	2,162	55
7	7	0.31	446	0.68	2,170	77
8	1	0.43	492	1.39	544	77
9	3	0.42	468	0.51	655	77
10	5	0.43	478	0.28	571	77
11	7	0.42	475	0.29	859	77

Pressure, equivalence ratio, inlet temperature, velocity, Reynolds number and nitrogen dilution. Cases with 77% vol. N_2 pertain to C_3H_8/air mixtures

The inlet Reynolds numbers, Re_{IN}, (based on the uniform inlet properties and the channel hydraulic diameter) were as high as 2171, leading to laminar flows; recent turbulent catalytic combustion studies [7] have shown that the strong flow laminarization induced by the hot catalytic walls guarantees laminar flow conditions at considerably higher Re_{IN}. Cases 1–7, characterized by the absence of homogeneous ignition, provided the platform for evaluating the catalytic reactivity and its pressure dependence with the aid of the Raman measurements. In Cases 8–11, whereby flames were established inside the channel, the gas-phase combustion processes were evaluated using the OH-LIF data.

4.3.1 Effect of Pressure on the Catalytic Reactivity

Comparisons between Raman-measured and predicted transverse profiles of C_3H_8 and H_2O mole fractions, at four selected streamwise locations, are illustrated in Fig. 4.1 for Cases 1–4; for clarity, 27 of the total 63 measured points are provided. Modest profile asymmetries are evident due to small differences between the upper- and lower-wall temperatures (see the wall temperatures of Fig. 4.2).

The experimentally-resolved transverse extent ranged from $0.3 \leq y \leq 6.7$ mm at high pressures down to $0.6 \leq y \leq 6.4$ mm for the atmospheric pressure experiments (due to the increasing Raman signal-to-noise ratio with rising pressure). The measurement accuracy was $\pm 10\%$ for compositions as low as 0.5% vol.; concentrations less than 0.5% vol. entailed larger measurement uncertainties. To further increase the Raman signal of the deficient reactant (propane), C_3H_8/air/O_2 mixtures were also examined in certain cases, whereby the volumetric content of nitrogen was as low as 55%.

Axial profiles of the computed catalytic (*C*) and gas-phase (*G*) propane conversion rates (the latter integrated over the 7-mm channel height), are presented in Fig. 4.2 for Cases 1–4.

The maximum reactor extent over which the *G* conversion is negligible has been delineated with the arrows in Fig. 4.2 (defined as the positions where the *G* conversion amounts to 5% of the *C* conversion). For the assessment of the catalytic reactivity, the forthcoming comparisons of Raman data with predictions are limited to this extent so as to avoid falsification of the catalytic kinetics by gas-phase chemistry. The initial reactor extent over which the gas-phase contribution could be safely neglected ranged from 10.9 cm (Case 3) to 14.2 cm (Case 4).

The heterogeneous reactivity of propane was assessed by varying the reactor pressure and surface temperature, and monitoring the near-wall bending of the propane transverse profiles. In Case 2 (Fig. 4.1, $p = 3$ bar) the catalytic fuel conversion is already appreciable at surface temperatures $T_w \leq 850$ K, while mass-transport-limited conditions are approached for $T_w = 1064$ K (manifested by the low propane mole fractions near both walls in Fig. 4.1(2d)).

In contrast, previous CH_4/air studies [8] in the same reactor necessitated an increase in pressure to achieve a similar level of reactivity under the same mass

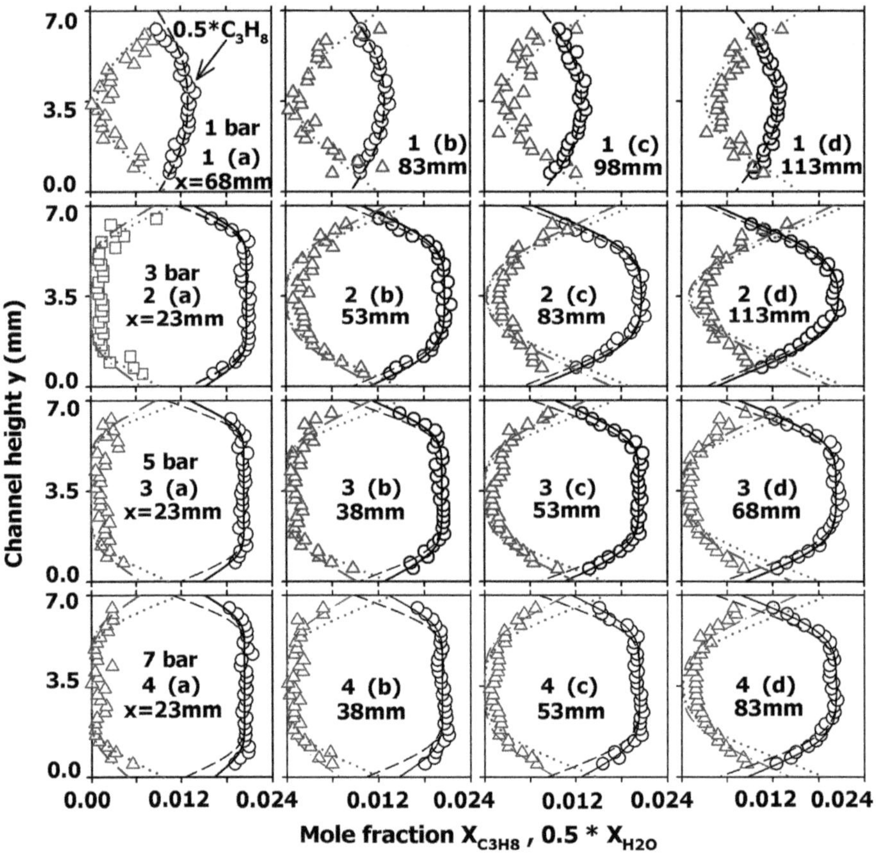

Fig. 4.1 Measured and predicted transverse profiles of C_3H_8 and H_2O mole fractions for Cases 1–4 at four streamwise locations. Measurements: C_3H_8 (*circles*), H_2O (*triangles*). Predictions: C_3H_8 (*solid lines*, pressure-corrected model; *dashed lines*, Garetto), H_2O (*dashed-dotted lines*, pressure-corrected model; *dotted lines*, Garetto)

inflow conditions. Numerical predictions using the global catalytic reaction rate of Garetto yielded a very good agreement with the measured C_3H_8 and H_2O profiles at pressures up to 3 bar (Fig. 4.1). Moreover, the increase in catalytic reactivity with increasing downstream surface temperatures and the approach to transport-limited conversion is also well captured at this pressure range (see Fig. 4.1(1a) through (1d)).

At pressures above 3 bar, the employed catalytic model could not capture the measured boundary-layer profile of propane (see Fig. 4.1(3, 4)); therein, the predictions clearly overpredicted the measured catalytic reactivity. This is attributed to the overall reaction order (and hence pressure order) of 1.15 (Eq. 4.1).

Recent catalytic combustion studies of methane on Pt [8] have shown that the heterogeneous reactivity of methane follows a $\sim p^{0.47}$ pressure dependence, over the range $1 < p < 16$ bar; this power law bears the combined effects of the first-order reaction dependence with respect to methane, and the $p^{-0.53}$ dependence due

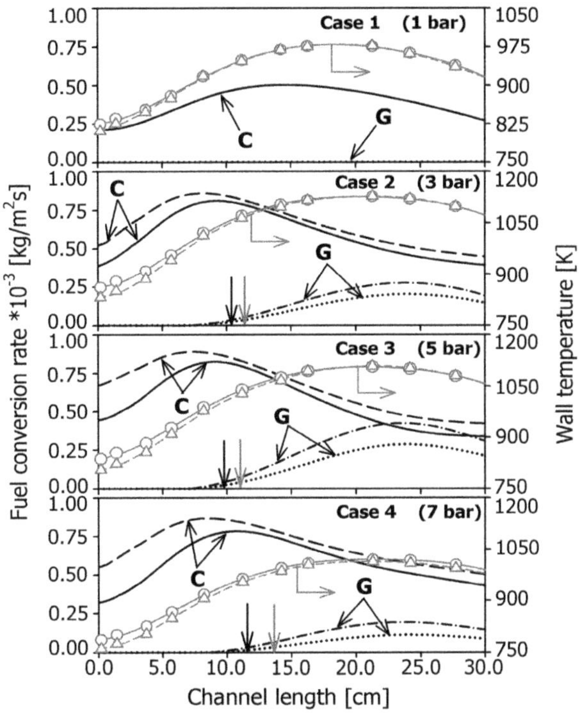

Fig. 4.2 Computed streamwise catalytic (*C*) and gas-phase (*G*) fuel conversion rates, averaged over *y* (*solid, dashed-dotted-lines*: pressure-corrected model; *dashed, dotted-lines*: Garetto). *Upper-* and *lower-wall* (*solid* and *dashed gray lines* respectively) temperature profiles fitted through thermocouple measurements (upper: *circles*; lower: *triangles*). The *vertical arrows* on the *x*-axes define the position of non-negligible gas-phase conversion (*black*: pressure-corrected model; *gray*: Garetto)

to the reduction of free platinum sites with increasing oxygen partial pressure. The latter mechanism is very important in restraining the rate of increase of the catalytic reactivity with rising pressure.

The challenge in a global catalytic step is to reproduce the reactivity over a wide range of operating conditions (pressures and temperatures). Appropriate pressure corrections have thus been proposed for global catalytic steps of methane [8]; with such corrections, the performance of single steps has been shown to be modest over the wide pressure range 1–16 bar.

In a fashion similar to the earlier methane studies, the current global propane step is extended by introducing a pressure term which restrains the rate of increase of the catalytic reactivity with increasing pressure:

$$\dot{s}_{C_3H_8} = A \times \left(\frac{p}{p_0}\right)^{-n} \times T_W^{1.15} \times \exp(-E_a/RT_W) \times [C_3H_8]_W^a \tag{4.2}$$

with n a positive number smaller than unity and $p_0 = 1$ bar. A value of $n = 0.4 \pm 0.03$ was found to yield the best fit to the experimentally acquired boundary layer propane profiles over the entire pressure range $1 \leq p \leq 7$ bar for all the cases considered in this work (Fig. 4.1).

Transverse profiles of major species were computed anew for Cases 2–4 with the proposed kinetics of Eq. (4.2) at locations upstream of considerable gas-phase fuel conversion. Significantly better agreement with Raman data was achieved for the higher pressure Cases 3 and 4 (5 and 7 bar respectively) as seen in Fig. 4.1, particularly for the lower examined surface temperatures. At the same time, the good model performance at pressures $p \leq 3$ bar was retained.

The large overprediction of catalytic reactivity by Eq. (4.1) is further evident in Fig. 4.2. For Case 4 (7 bar), the fuel conversion rate is overpredicted by as much as 42% for a surface temperature of 786 K, which in turn yields the same overprediction in heterogeneously-produced water (Fig. 4.1(4)). At higher wall temperatures, however, mass-transport-limited operation is approached and both models yield similar results: the difference between both model predictions drops

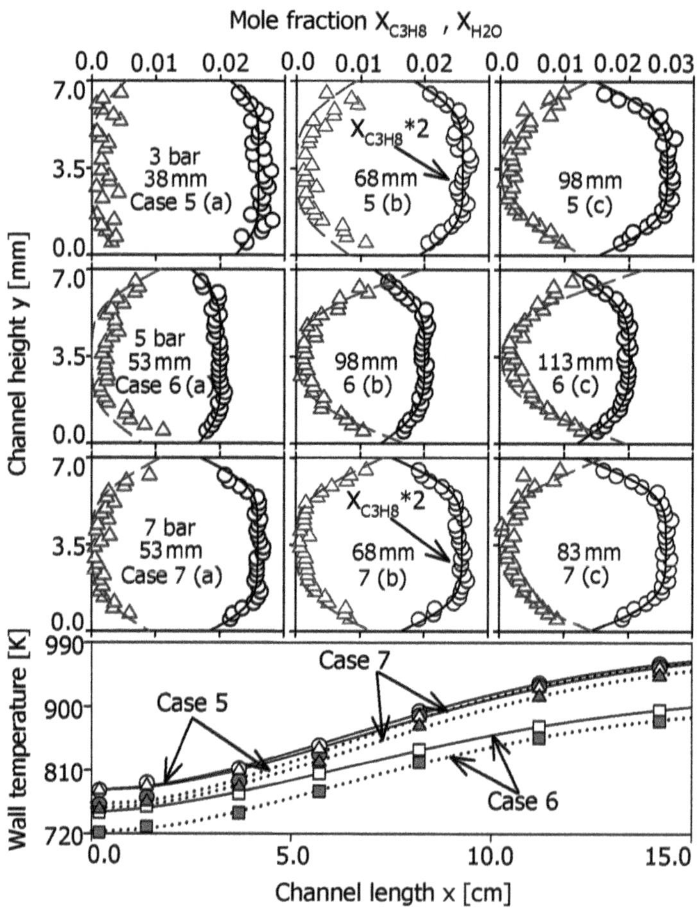

Fig. 4.3 Measured and predicted transverse profiles of C_3H_8 and H_2O mole fractions for Cases 5, 6 and 7 at three streamwise locations. Measurements: C_3H_8 (*circles*), H_2O (*triangles*). Predictions: C_3H_8, *solid lines*; H_2O, *dashed-dotted lines*. For clarity, the scale of propane has been expanded by a factor of 2 in Cases 5 and 7. *Upper-* and *lower-wall* (*solid* and *dotted-lines* respectively) temperature profiles fitted through thermocouple measurements (upper: *open symbols*; lower: *gray-filled symbols*) for Cases 5 (*circles*), 6 (*squares*) and 7 (*triangles*)

to 5.2% at 995 K. It is noted that the enhanced performance of Eq. (4.2) yielded a more conservative estimate of the reactor extent with minimal gas-phase contribution. As seen in Fig. 4.2, the computed lengths of significant gas-phase conversion are reduced by as much as 13 and 21 mm for Cases 3 and 4 respectively. The associated higher near-wall fuel concentration, along with the increased concentration of catalytically produced water in the gas-phase induction zone, have been shown to affect on the onset of homogeneous ignition [6].

The applicability of the pressure-corrected global step was further evaluated by comparing numerically-predicted species profiles with Raman data for Cases 5–7 (Fig. 4.3). The propane and water profiles are captured correctly at all pressures and for all surface temperatures considered. Cases 5 and 7 pertain to propane/air mixtures (without oxygen addition) with same equivalence ratio and mass inflow and similar surface temperatures, at 3 and 7 bar respectively. An increased catalytic reactivity from 3 to 7 bar is evident by comparing propane profiles at position $x = 68$ mm, with upper wall temperatures of 864 K and 868 K for Cases 5 and 7 respectively. Case 6 is provided here to exemplify the applicability of the extended model at lower equivalence ratios ($\varphi = 0.23$) and lower wall temperatures (723 K $\leq T_{\mathrm{wall}} \leq$ 898 K).

4.3.2 Homogeneous Ignition

It can be argued that a global catalytic step that captures (over a prescribed range of operating parameters) the heterogeneous fuel consumption, when used in conjunction with a detailed homogeneous reaction scheme, can also reproduce the onset of homogeneous ignition. Earlier catalytic combustion studies of hydrogen and methane [6, 9] with detailed hetero-/homogeneous reaction schemes have clarified the particularly weak coupling between the two pathways via radical adsorption/desorption reactions; this coupling had a minimal effect on the radical pool over the gas-phase induction zone [6, 9, 10]. Major hetero-/homogeneous interactions come from the near-wall catalytic fuel depletion, which in turn inhibits homogeneous ignition [11], and a secondary chemical coupling originating from the heterogeneously-produced major species (notably H_2O), which impact homogeneous ignition (promoting for methane [12, 13], inhibiting for hydrogen [9, 14]).

These two effects can be captured by a global catalytic step. Coupling via other intermediate species, such as CO, can be important [8, 15] in determining the overall consumption of this species and in determining the resulting CO emissions but its impact on homogeneous ignition of methane is secondary [12]. Previous hydrogen and methane homogeneous ignition studies have clearly pointed out that deficiencies in homogeneous ignition predictions stem mostly from low-temperature gas-phase kinetics and not from heterogeneous kinetics [9, 12].

Comparisons between LIF-measured and numerically predicted distributions of the OH radical are illustrated in Fig. 4.4 for Cases 8–11 of Table 4.1. In all cases, the pressure-corrected heterogeneous model established in the previous section

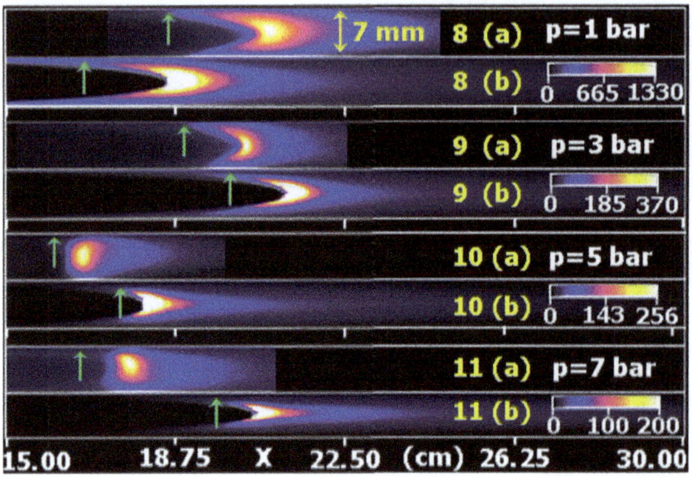

Fig. 4.4 Measured and predicted distributions of the OH radical for Cases 8–11. (a) OH-LIF, (b) numerical predictions with the Qin et al. (gas-phase) and pressure-corrected (catalytic) reaction schemes. The *green arrows* define the onset of homogeneous ignition. Predicted OH levels in ppmv

was used, coupled to the optimized gas-phase mechanism of Qin et al. The slight asymmetries of the flames are attributed to temperature differences between the two catalytic walls. The location of homogeneous ignition (χ_{ig}), shown with the green arrows in Fig. 4.4, was determined in both experiments and predictions as the streamwise position where OH levels rose to 5% of their maximum flame values. The OH color-coded bars (ppmv) in Fig. 4.4 refer to the predictions (the OH-LIF measurements were in the linear regime but absolute concentrations were not deduced). The lower pressure flames exhibited the highest absolute OH levels, which dropped rapidly with increasing pressure.

Having established the applicability of the pressure-corrected global step in the previous section, the comparisons in Fig. 4.4 allow for a direct evaluation of the gas-phase reaction scheme in predicting ignition delay times and laminar flame speeds (linked to the homogeneous ignition distances and to the flame sweep angles inside the channel, respectively). Figure 4.4 indicates a moderate ability of the gaseous scheme in capturing correctly the measured flame shapes, with better agreement at lower pressures.

However, of greater importance is the ability of the gaseous scheme to correctly predict homogeneous ignition distances (since homogeneous ignition is deemed as detrimental to the reactor integrity). An overall good agreement is achieved between measurements and predictions; the ignition distance is underpredicted by 9.3% for Case 8 (1 bar), while pressures up to 5 bar consistently yielded overpredictions (by up to 9.8% for Case 10). Increasing the pressure to 7 bar resulted in greater overpredictions of χ_{ig} (by 19% for Case 11). These observations are in accordance with recent numerical studies that exemplified the gas-phase

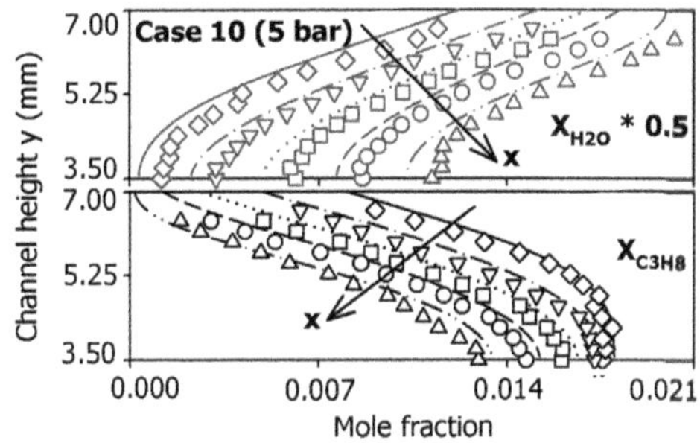

Fig. 4.5 Predicted (*lines*) and measured (*symbols*) profiles over the channel half-height of C_3H_8 and H_2O mole fractions for Case 10 at streamwise positions $x = 23$ (*solid lines*, diamonds), 53 (*dashed-dotted lines, lower triangles*), 83 (*dotted lines, squares*), 113 (*dashed lines, circles*) and 143 (*dashed-doubled-dotted lines, upper triangles*) mm

mechanism's inability to correctly capture experimentally observed ignition delay times for propane at moderate-to-high pressures ($p > 4$ bar) [16].

Further validation for the applicability of the coupled hetero-/homogeneous models is provided in Fig. 4.5; Raman-measured major species profiles over the channel half-height are compared against numerical predictions obtained using the pressure-corrected global step and the gas-phase scheme, for selected streamwise positions of Case 10 prior to homogeneous ignition. The predictions are in good agreement with the experiments, indicating that the chemical models capture correctly the fuel consumption in the reactor.

In Fig. 4.6, catalytic and gas-phase conversion rates along with propane and OH radical mole fractions are plotted for Case 9. It is evident that gas-phase chemistry is responsible for most of the fuel conversion, well before the point of

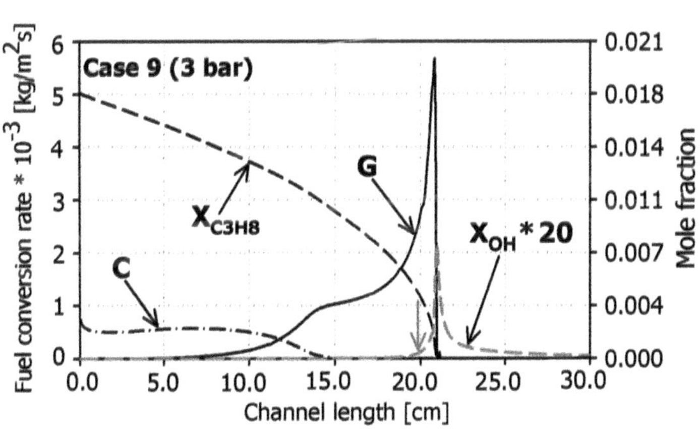

Fig. 4.6 Predicted catalytic (*C*) and gas-phase (*G*) conversion rates for the entire channel length for Case 9, along with propane and OH radical species mole fractions. The vertical grey arrow on the *x*-axis indicates the predicted point of homogeneous ignition

homogeneous ignition. Specifically, pre-ignition gas-phase chemistry already surpasses catalytic chemistry in terms of fuel conversion at $x = 11.8$, where only $\sim 32\%$ of the fuel has been converted.

The significant amount of fuel available for gas-phase reactions is attributed to the small surface-to-volume ratio of the channel, typical of the optically accessible reactor, along with the larger-than-unity Lewis number of propane, which hinder propane from reaching the catalytic surfaces fast enough.

4.4 Conclusions

The pure heterogeneous and the coupled hetero-/homogeneous combustion of lean propane/air mixtures over platinum was investigated experimentally in an optically accessible, channel flow reactor at pressures 1 bar $\leq p \leq$ 7 bar with in situ, 1-D Raman and planar OH-LIF measurements. The catalytic fuel conversion rate was assessed as a function of pressure, equivalence ratio and surface temperature. A global-step catalytic reaction for propane was established, which correctly accounted for the pressure dependence of the catalytic reactivity over the entire investigated pressure range. Numerical predictions using a detailed homogeneous reaction scheme coupled with the established catalytic reaction step reproduced the onset of gas-phase ignition at moderate pressures ($p \leq 5$ bar), with somewhat more pronounced overpredictions at higher pressures (7 bar). The overall performance of the employed hetero-/homogeneous reaction models verified their applicability in the design of propane-fueled catalytic microreactors.

References

1. Garetto TF, Rincon E, Apesteguia CR (2004) Deep oxidation of propane on Pt-supported catalysts: drastic turnover rate enhancement using zeolite supports. Appl Catal B: Environ 48:167–174
2. Qin Z, Lissianski VV, Yang H, Gardiner WC, Davis SG, Wang H (2000) Combustion chemistry of propane: a case study of detailed reaction mechanism optimization. Proc Combust Inst 28:1663–1669
3. Coltrin ME, Kee RJ, Rupley FM, Chemkin S (1996) A Fortran package for analyzing heterogeneous chemical kinetics at the solid surface-gas phase interface, Report No. SAND90-8003C; Sandia National Laboratories
4. Kee RJ, Rupley FM, Miller JA, Chemkin II (1996) A Fortran chemical kinetics package for the analysis of gas-phase chemical kinetics, Report No. SAND89-8009B; Sandia National Laboratories
5. Kee RJ, Dixon-Lewis G, Warnatz J, Coltrin ME, Miller JA (1996) A Fortran computer code package for the evaluation of gas-phase multicomponent transport properties, Report No. SAND86-8246; Sandia National Laboratories
6. Reinke M, Mantzaras J, Schaeren R, Bombach R, Inauen A, Schenker S (2005) Homogeneous ignition of CH_4/air and H_2O- and CO_2-diluted CH_4/O_2 mixtures over platinum; an experimental and numerical investigation at pressures up to 16 bar. Proc Combust Instit 30:2519–2527

7. Appel C, Mantzaras J, Schaeren R, Bombach R, Inauen A (2005) Turbulent catalytically stabilized combustion of hydrogen/air mixtures in entry channel flows. Combust Flame 140:70–92
8. Reinke M, Mantzaras J, Schaeren R, Bombach R, Inauen A, Schenker S (2004) High-pressure catalytic combustion of methane over platinum: in situ experiments and detailed numerical predictions. Combust Flame 136:217–240
9. Appel C, Mantzaras J, Schaeren R, Bombach R, Inauen A, Kaeppeli B, Hemmerling B, Stampanoni A (2002) An experimental and numerical investigation of homogeneous ignition in catalytically stabilized combustion of hydrogen/air mixtures over platinum. Combust Flame 128:340–368
10. Reinke M, Mantzaras J, Schaeren R, Bombach R, Kreutner W, Inauen A (2002) Homogeneous ignition in high-pressure combustion of methane/air over platinum: comparison of measurements and detailed numerical predictions. Proc Combust Inst 29:1021–1029
11. Mantzaras J (2006) In: Jiang SZ (ed) Focus on combustion research. Nova Publishers, New York, p 241
12. Reinke M, Mantzaras J, Bombach R, Schenker S, Inauen A (2005) Gas-phase chemistry in catalytic combustion of methane/air mixtures over platinum at pressures of 1 bar to 16 bar. Combust Flame 141:448–468
13. Reinke M, Mantzaras J, Bombach R, Schenker S, Tylli N, Boulouchos K (2007) Effects of H_2O and CO_2 dilution on the catalytic and gas phase combustion of methane over platinum at elevated pressures. Combust Sci Technol 179:553–600
14. Bui PA, Vlachos DG, Westmoreland PR (1996) Homogeneous ignition of hydrogen/air mixtures over platinum. Proc Combust Inst 26:1763–1770
15. Karagiannidis S, Mantzaras J, Jackson G, Boulouchos K (2007) Hetero-/homogeneous combustion and stability maps in methane-fueled catalytic microreactors. Proc Combust Inst 31:3309–3317
16. Jomaas G, Zheng XL, Zhu DL, Law CK (2005) Experimental determination of counterflow ignition temperatures and laminar flame speeds of C2–C3 hydrocarbons at atmospheric and elevated pressures. Proc Combust Inst 30:193–200

Chapter 5
Experimental and Numerical Investigation of a Propane-Fueled, Catalytic, Mesoscale Combustor

5.1 Preface

The present work undertakes a combined experimental and numerical investigation of a propane-fueled, platinum-coated, mesoscale combustor, which is candidate for portable power generation applications. As a first step, the heterogeneous reactivity of propane on platinum has been assessed with gas-phase Raman spectroscopy and an appropriate global reaction step for lean propane/air combustion was constructed, valid at pressures $1 \leq p \leq 7$ bar, a range of interest to microreactors and small-scale industrial turbines. In a second step, a detailed numerical parametric study in a single catalytic channel has been performed, with inlet conditions and combustor material properties as the parameters of main interest, aimed at defining regimes of optimal power output and fuel conversion. Detailed numerical treatment was provided for the heat transfer mechanisms inside the channel, which included heat conduction in the solid and surface radiation heat exchange between the reactor elements themselves and between the reactor and its surroundings. Based on the findings of the aforementioned parametric study, the third and final step involved the construction and testing of a mesoscale catalytic combustor for a range of operational parameters relevant to micro- and mesoscale power generation applications. A continuum model for the entire combustor monolith was developed to complement the experiments, thus allowing for the simulation of the temperature 2-D distribution inside the combustor. This numerical tool facilitated the detailed modeling of the heat losses from the catalytic combustor, evident in the experiments, and allowed for further design improvements.

S. Karagiannidis, *Catalytic Microreactors for Portable Power Generation*,
Springer Theses, DOI: 10.1007/978-3-642-17668-5_5,

5.2 Motivation and Background

The pursuit of a mesoscale catalytic combustor was motivated, as mentioned in Chap. 1, by the initiative to construct a small-scale, gas turbine-driven power generation unit in the order of a few hundred W_{el} (electric) [1]. A demonstration unit is developed for this purpose, with the catalytic combustor specifications dictating a power output of $P_{OUT} = 9{,}400\ W_{th}$, a mass flow rate of $m_{IN} = 19.5$ g/s, an estimated combustor volume of $\sim 200\ cm^3$ and a corresponding gas hourly space velocity (GHSV) in the order of $10^5\ h^{-1}$. The thermal cycle analysis of the mesoscale gas-turbine unit also dictated a turbine inlet temperature and a target pressure ratio at the compressor discharge of $\sim 1{,}150$ K and 2.80, respectively. Catalytic combustion at the mesoscale offers, apart from the benefits already mentioned in Sect. 1.2, additional advantages in simplified combustor design and manufacturing. Moreover, due to the inherent constraints of small-scale turbines (e.g. inability to cool the turbine blades) the use of a catalytic monolithic reactor is particularly advantageous as it allows for relatively low outlet temperatures and low pressure drops. Scaling of such a combustor is simplified by the fact that, once a single channel of the monolith is studied, power output can be adjusted by evaluating the number of channels required in the monolithic structure. An overall system efficiency of $\sim 10\%$ is considered, the rule of thumb being that the combustor should have a thermal power roughly ten times the desired electrical power. In order to realize the mesoscale gas-turbine unit, a number of different approaches have been investigated, with the single-stage, recuperated thermal cycle concept chosen not only for ease of design and manufacturing, but also for simplification in the subsequent numerical modeling of the individual components.

5.3 Experimental Setup

The test rig employed in this study consisted of a high-pressure cylindrical steel vessel with a length of 1.8 m and an internal diameter of 0.28 m (see Chap. 2). Visual inspection and accessibility of the reactor assembly was achieved via a 50 mm diameter quartz window at the rear flange of the vessel and two 350 mm long and 50 mm high quartz windows at the vessel sides. Experiments were performed in two different reactors. The first reactor was the optically accessible, channel-flow reactor presented in Sect. 2.1; therein, the use of in situ laser diagnostics (Sect. 2.2) allowed for the assessment of the catalytic reactivity of propane on platinum. The second reactor was a honeycomb subscale unit of the desired catalytic combustor (see Sect. 2.3), and was used for the evaluation of combustor performance characteristics.

5.4 Numerical Model

5.4.1 Single Channel Parametric Simulations

The full-elliptic 2-D laminar CFD code presented in Sect. 3.1 was used to simulate a plane channel-flow configuration, with channel length $L = 75$ mm, height $2b = 1$ mm and wall thicknesses $\delta = 0.1$ mm. Due to the small hydraulic diameter of the channel and strong laminarization of the flow at the high wall temperature regions, laminar conditions are maintained throughout the channel, even at conditions of high inlet velocity [2]. For the heat conduction in the solid material, a 2-D approach is also adopted with the solid thermal conductivity being either $k_s = 2$ or 14 W/mK. The outer channel wall surfaces were treated as adiabatic, thus resembling an inner channel of the catalytic monolithic combustor. Parametric studies were carried out by varying the inlet velocity U_{IN} and the fuel-to-air equivalence ratio φ. The inlet pressure was fixed at $p = 2.5$ bar, while the inlet mixture preheat was $T_{IN} = 750$ K, considering recuperated operation.

Radiation heat transfer exchange between the discretized catalytic surface elements as well as radiation losses from each internal surface element to the reactor inlet and outlet areas were accounted for by the net radiation method for diffuse-gray areas [3, 4]. The inlet, outlet and channel element emissivities were all equal to $\varepsilon = 0.6$. Gas radiative emission and absorption was neglected given the small optical paths and the large nitrogen dilution in the feed. Radiative boundary conditions were applied at the inlet and outlet vertical wall faces, allowing for heat losses towards the surroundings.

Gas-phase chemistry has been shown to affect combustion characteristics in catalytic reactors, even in sub-quenching channel confinements and particularly at elevated pressures [5]. To allow for extended parametric studies, gas-phase chemistry has not been considered. In doing so, a conservative estimate of the stable combustion regimes was obtained, since gas-phase chemistry has been shown to extend the stability limits of catalytic combustors [3]. Finally, mixture-average diffusion provided the gas-phase transport model [6].

An orthogonal staggered mesh of 24 × 200 points (in the x- and y-direction, respectively) over half the gas-phase domain with finer spacing near the catalytic walls yielded a grid-independent solution. Two-dimensional solid heat conduction was solved on a uniform grid of 20 × 200 points. Uniform inlet properties were used for temperature, axial velocity and species mass fractions. No-slip was applied for both velocity components at the gas-wall interface ($y = b$) and zero-Neumann conditions were set at the outlet ($x = L$) and the plane of symmetry ($y = 0$).

5.4.2 Monolith Continuum Model

In the continuum model approach, the two-dimensional transient energy equation under the finite volume approach for the entire monolithic structure was solved,

Table 5.1 Material properties of the catalytic monolith (FeCr alloy wall and porous Al_2O_3 washcoat properties were taken into account)

Heat conduction $k_{c,x}$ (axial direction) (W/mK)	1.80
Heat conduction $k_{c,r}$ (radial direction) (W/mK)	0.20
Density ρ_c (kg/m^3)	1,253
Heat capacity c_c (J/kgK)	620

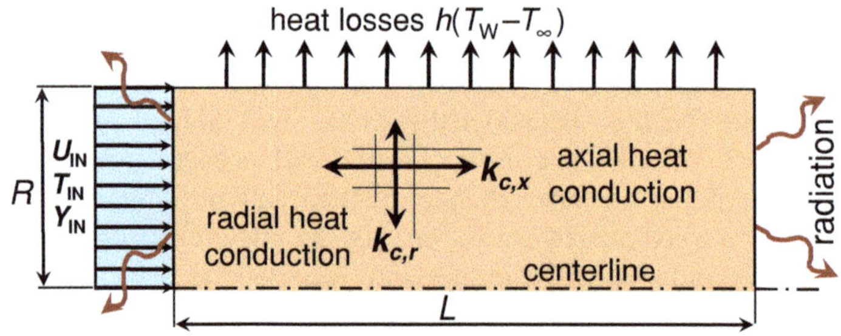

Fig. 5.1 Schematic of the continuum model for the combustor monolith structure

as described in Sect. 3.2, with monolith density ρ_c, heat capacity c_c and thermal conductivity k_c weighted functions of the solid material and gas phase. Table 5.1 provides the values used in this work, calculated using the properties of the combustor foils, the catalytic washcoat and the propane/air mixture at the corresponding experimental conditions. Variable thermal conductivity in the axial and radial directions was also accounted for [7].

The subscale reactor used in the experiments was modeled with a length of $L = 75$ mm and a radius or $R = 17.5$ mm, using a uniform grid of 200×35 points. The coupling to the reactive flow fields of the catalytic channels in the monolith was achieved with the use of the heat source term q, calculated for a representative number of channels, which accounted for the heat transfer between gas phase and solid by the change in the integral enthalpy flux in the gas phase. In this work, the flow fields of 8 equidistant channels along the combustor monolith radius were computed. The entire honeycomb structure had $\sim$650 channels. It is noted that even though the time integration if Eq. 3.19 provided the transient response of the reactor, of interest in this work is only the steady-state solution wherein experiments were available. Heat losses due to radiation at the monolith inlet and outlet, along with convective heat losses from its outer wall surface, were included as boundary conditions. In the model, inlet and outlet enclosures were treated as black bodies with temperatures equal to the inlet and outlet gas temperatures, respectively (values taken directly from the experiments), while the convective heat transfer coefficient was varied to match the experimentally observed exhaust gas temperatures (see Fig. 5.1). Surface reaction rates were evaluated with Surface-CHEMKIN [8] and mixture-average diffusion was the transport model, using the CHEMKIN transport database [6].

5.5 Results and Discussion

Recent catalytic combustion studies have shown that the heterogeneous reactivity of hydrocarbons on platinum follows a positive power law pressure dependence, which bears the combined effects of a positive, first-order reaction dependence with respect to fuel concentration in the gas phase over the catalyst surface, and a negative dependence due to the reduction of free platinum sites with increasing oxygen partial pressure. The latter mechanism is very important in restraining the rate of increase of the catalytic reactivity with rising pressure. The challenge in a global catalytic step is to reproduce the reactivity over a wide range of operating conditions (pressures and temperatures). As presented already in Chap. 4, the global step of Garetto et al. [9] for propane oxidation on platinum at atmospheric pressure was extended over the pressure range $1 \leq p \leq 7$ bar by introducing a pressure correction term which restrains the rate of increase of the catalytic reactivity with rising pressure [10]. Figure 5.2 provides a comparison between numerically predicted (using the extended global step) and experimentally observed (Raman data) transverse profiles of propane and water mole fractions over the transverse distance of the channel-flow reactor at selected axial locations for two experimental cases. The good agreement between experimental and numerical results evident in Fig. 5.2 confirms the validity of the pressure-corrected global step for describing lean propane/air combustion on platinum. It is noted here that experiments presented in Fig. 5.2 where performed using propane/air (instead of propane/air/oxygen) mixtures and were used to validate the extended global step under real application conditions.

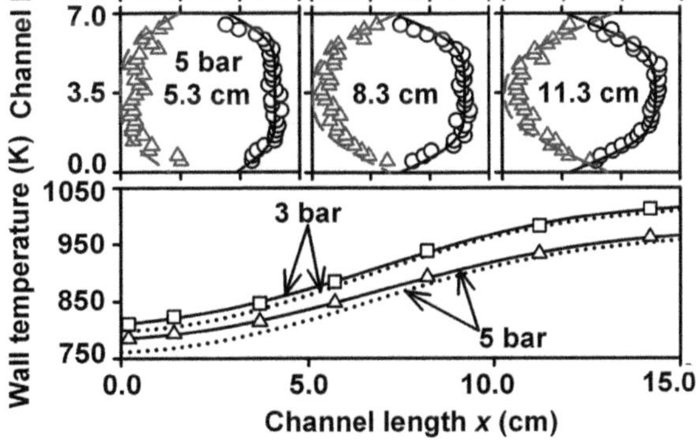

Fig. 5.2 Measured and predicted transverse profiles of C_3H_8 and H_2O mole fractions at three streamwise locations for $T_{IN} = 449$ K, $\varphi = 0.31$, $U_{IN} = 1.61$ and 0.96 m/s, at 3 and 5 bar, respectively. Measurements: C_3H_8 (*circles*), H_2O (*triangles*). Predictions: C_3H_8, *solid lines*; H_2O, *dashed-dotted lines*. Upper- and lower-wall (*solid* and *dotted-lines*, respectively) temperature profiles fitted through thermocouple measurements (*symbols* indicate thermocouple positions)

5.5.1 Single-Channel Parametric Study

In the ensuing calculations, power outputs were computed and combustion efficiencies were estimated. The wall thermal conductivity was set to $k_s = 2.0$ or 14 W/mK, thus simulating a ceramic and a metallic reactor (those values roughly correspond to cordierite and FeCr alloy materials, respectively). Mixture compression was fixed at $p = 2.5$ bar, the equivalence ratios were $\varphi = 0.40, 0.35$ and 0.33, while the channel wall thickness was $\delta = 0.1$ mm. The forthcoming diagrams were constructed by one-parameter continuation of the mixture inlet velocity U_{IN}, which was increased until a critical value was reached with a marked drop in reactor performance. The main objective was to construct power curves for a given set of reactor geometry, properties, and inlet conditions. An example of the results obtained can be seen in Fig. 5.3, where temperature and fuel distributions

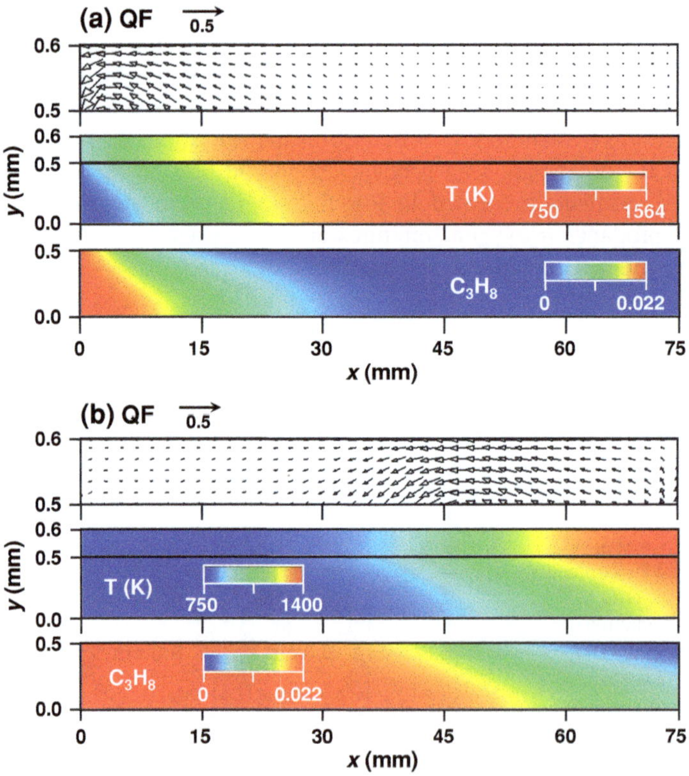

Fig. 5.3 Two-dimensional distributions of temperature and propane mass fraction and heat flux vector plot in the solid (*top*, *middle* and *bottom panels*, respectively) for two cases with $\varphi = 0.35$, $k_s = 14$ W/mK and **a** $U_{\mathrm{IN}} = 2.3$ m/s, **b** $U_{\mathrm{IN}} = 4.5$ m/s. Reference vectors denote heat flux magnitude of 0.5 W/mm^2. The plane of symmetry is at $y = 0$ and the gas–solid interface at $y = 0.5$ mm

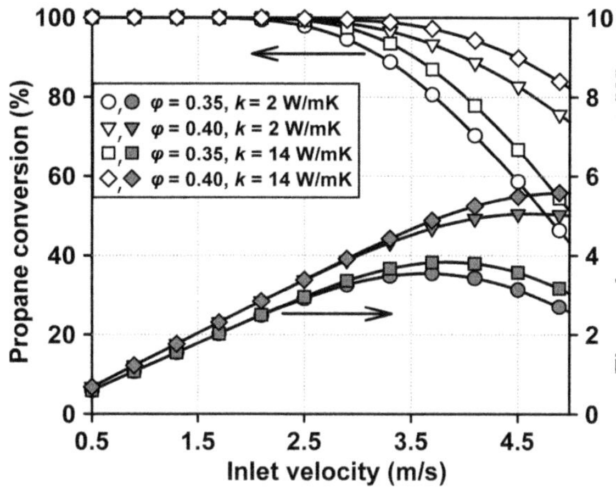

Fig. 5.4 Fuel conversion (*open symbols*) and thermal power output (*closed symbols*) versus inlet velocity. Thermal power output is computed for a channel of 1 mm^2 cross-section

inside the channel are provided for two cases with equivalence ratio $\varphi = 0.35$, thermal conductivity $k_s = 14$ W/mK and two inlet velocities $U_{IN} = 2.3$ m/s and $U_{IN} = 4.5$ m/s. As will become evident later in Fig. 5.4, both cases exhibited similar thermal power outputs but differed substantially in overall propane conversion efficiency. The heat flux inside the solid wall is visualized with the help of vector plots. It is clarified that the transverse temperature differences inside the solid were very small (up to 1.5 K over the entire 0.1 mm channel wall thickness).

Thermal power curves and combustion efficiencies for a reactor of length $L = 75$ mm with a channel cross-section of 1×1 mm^2 are summarized in Fig. 5.4. The velocity ranges from $U_{IN} = 0.5$ m/s, set as initial velocity for all cases studied in this work, to 5.0 m/s; the latter is arbitrarily chosen as a representative value where the combustion efficiency (defined as the temperature rise of the gas at the channel outlet over the potential adiabatic temperature gain) drops significantly.

For moderate inlet velocities, which can reach up to 2.5 m/s for the conditions of Fig. 5.4, a linear increase in power output is observed with rising inlet velocity. The combustion efficiency remains high since almost complete fuel conversion is achieved (~99.99%), while the power output is directly proportional to the equivalence ratio. As long as fuel conversion is nearly complete, no difference is observed between reactors of different solid thermal conductivities. The linearity in reactor response ceases as the inlet velocity approaches a critical value. With increasing inlet velocity, every power curve reaches a maximum. These maxima lie at combustion efficiencies (i.e. fuel conversions) of 71–83% for all cases considered.

The effect of wall thermal conductivity is pronounced at high inlet velocities, where increased efficiency is observed in the case of high solid thermal conductivity, as seen in Fig. 5.5 pertaining to $U_{IN} = 3.7$ m/s. In this graph, axial profiles of the energy heat balance terms in the solid are provided in terms of all modeled heat transfer modes: heat generated via reactions on the surface, heat convected to the

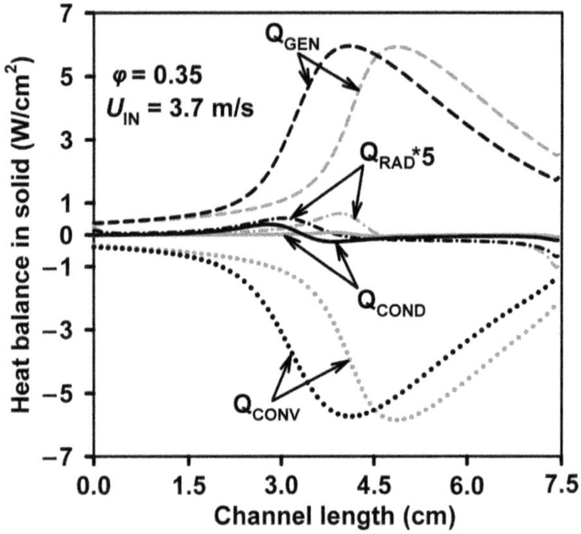

Fig. 5.5 Axial heat balance in the solid wall for $k_s = 2$ W/mK (*grey curves*) and 14 W/mK (*black curves*). *Dashed curves*: heat generated from catalytic reactions. *Dotted curves*: heat convected to the gas. *Solid curves*: heat conducted in the solid. *Dash-dotted curves*: surface radiation heat exchange

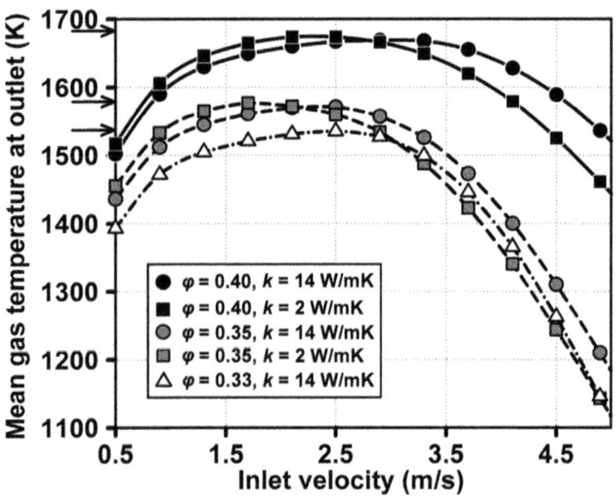

Fig. 5.6 Mean gas temperature at the channel outlet versus inlet velocity for three equivalence ratios and thermal conductivities k_s of 2 and 14 W/mK. The *horizontal arrows* on the *y* axis denote adiabatic temperatures for $\varphi = 0.33$, 0.35 and 0.40 at $T_{ad} = 1{,}537$, 1,579 and 1,683 K, respectively

gas, heat conducted in the solid wall and net radiation heat exchange. The radiation exchange contribution is altogether minimal and for clarity it has been multiplied by a factor of five.

A small amount of heat is conducted inside the solid wall for $k_s = 14$ W/mK (the corresponding amount for $k_s = 2$ W/mK is negligible). In the former case, the amount of heat feedback to the reactor front via solid conduction is much smaller compared to the maximum heat generation; nonetheless, this small amount is crucial in reducing the light-off distance and thus increasing the combustion efficiency.

Another interesting observation is presented in Fig. 5.6. Average gas temperatures at the reactor outlet are plotted versus inlet velocity for the two different solid thermal conductivities and equivalence ratios of Fig. 5.4. For comparison purposes, average gas temperatures are also plotted for $k_s = 14$ W/mK and $\varphi = 0.33$. Over a wide range of inlet velocities (between ~0.7 and 2.9 m/s for $\varphi = 0.35$ and $\varphi = 0.33$), the gas outlet temperatures vary by less than 10% around a temperature that depends on equivalence ratio. Reactors with high wall thermal conductivity are shown to have more uniform outlet temperatures than reactors with low wall thermal conductivity, over a somewhat wider range of power outputs and inlet velocities. This holds also true for the high equivalence ratios considered in this work, whereby this behavior is more pronounced. In mesoscale applications, such as the one that motivated the present study, this particular response to varying inlet conditions becomes important, since thermal management issues, along with operating cycle adjustments, are more difficult to tackle at small scales.

At the low inlet velocity regime (up to ~1.5 m/s), although complete fuel conversion is achieved, the outlet gas temperatures deviate significantly from the adiabatic flame temperature for the given inlet conditions of mixture preheat and fuel/air equivalence ratio. Radiation exchange between the cold inlet section and the hot channel walls near the entry is proportional to the fourth power of temperature of these surfaces, while heat generation inside the channel is proportional to the mass inflow. As the equivalence ratio is increased, it becomes less favorable to operate at low inlet velocities, an observation which holds true for both high and low wall thermal conductivities, even though the higher residence times and wall temperatures favor complete fuel conversion. The associated high adiabatic temperature rise necessitates higher inlet velocities to compensate for the higher radiation heat losses.

5.5.2 Experimental Testing of Mesoscale Combustor

The foregoing parametric study of a single catalytic channel dictated the construction of a mesoscale combustor with metallic walls, operating at high inlet velocities and low equivalence ratios, with the dimensioning mentioned in the experimental section. The combustor was subsequently tested for the experimental conditions of Table 5.2.

For a given set of inlet pressure p, equivalence ratio φ and inlet temperature T_{IN}, the mass inflow (air and propane streams) was varied from a minimum of 2.3 g/s up to 6.6 g/s with increments of ~0.5 g/s. After each mass inflow increment, exhaust gas temperature measurements were recorded once the combustor

Table 5.2 Mesoscale catalytic combustor experimental conditions

Inlet pressure p_{IN} (bar)	1.2, 2, 3, 5
Inlet temperature T_{IN} (K)	604–750
Equivalence ratio φ	0.20, 0.25, 0.30, 0.33
Mass inflow m_{IN} (g/s)	Up to 6.6

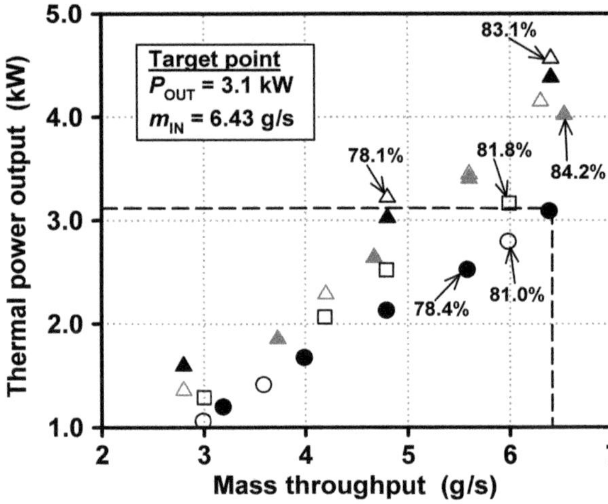

Fig. 5.7 Combustor thermal power output versus mass throughput deduced from measurements; $p_{IN} = 2$ and 3 bar (*closed* and *open symbols*), $T_{IN} = 600$ and 700 K (*black* and *gray symbols*) and $\varphi = 0.2, 0.25$ and 0.3 (*circles*, *squares* and *triangles*, respectively). For selected cases, combustor efficiencies are indicated on the *graph*. The target point for the subscale combustor was set at $P_{OUT} = 3.1$ kW at a mass inflow of $m_{IN} = 6.43$ g/s

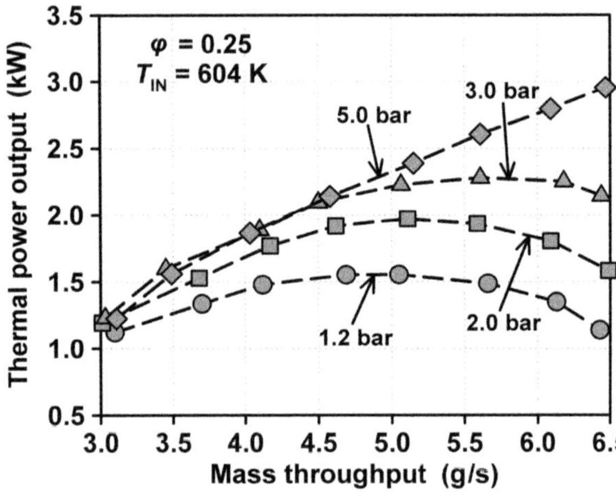

Fig. 5.8 Thermal power output versus mass throughput, measured for an equivalence ratio $\varphi = 0.25$ and inlet temperature $T_{IN} = 604$ K, over the investigated pressure range $1 \leq p \leq 5$ bar

reached steady state and the thermocouple readings were constant in time. Only heterogeneous combustion was considered; in a fashion similar to earlier studies [11], combustor efficiency was calculated by comparing the average temperature rise over the catalytic reactor $\Delta T = (1/2)(T_B + T_C)$, T_B and T_C being the exhaust gas thermocouple measurements at the combustor outlet, with the potential adiabatic temperature gain: $\eta = \Delta T/(T_{AD} - T_{IN})$; T_{AD} was calculated for each set p, φ and T_{IN} studied. Total thermal power output for the combustor was subsequently calculated based on the chemical energy input for the given set of inlet conditions and propane conversion.

Measured operational maps of thermal power output versus mass inflow, such as the one provided in Fig. 5.7, were constructed based on efficiency measurements for various sets of inlet pressure, temperature and mixture composition. These maps provided validation of the initial combustor design, since the target operational point of thermal power output and mass throughput, as dictated by the thermodynamic analysis of the proposed micro-gas-turbine system, was met or even exceeded by the combustor at various sets of inlet conditions.

Figure 5.8 provides power output curves measured for a fixed equivalence ratio and inlet temperature over the entire pressure range investigated. A significant rise in performance is evident as the inlet pressure is increased, with an almost threefold gain in power output for a fivefold increase in inlet pressure, at the maximum mass inflow value tested. It is emphasized that the correct kinetic expression with an appropriate pressure exponent derived in Eq. 4.2 is cardinal in assessing the effect of pressure.

A combination of reduced catalytic reactivity and short residence times at the low inlet pressures are, to a certain extent, responsible for the significant efficiency differences between operating pressures at constant mass inflows. However, for all cases considered in this work, the maximum combustor efficiency observed was ~85%, which indicates that heat losses to the surroundings and potential fuel breakthrough can further reduce the thermal power output of the combustor.

5.5.3 Continuum Model Simulations

Numerical simulations of the two-dimensional temperature distribution inside the catalytic monolith were performed using the continuum model approach described earlier, in order to facilitate the experiments conducted with the subscale model by providing insight on thermal management issues of the mesoscale combustor (namely radiation losses and heat losses from the outer honeycomb surface). The extended experimental investigation allowed only a selected number of operating conditions to be simulated. By varying only the external heat losses coefficient at the outer reactor surface, the computed outlet gas temperature of the monolith was matched with the exhaust gas thermocouple measurements at the combustor outlet. In all cases, good agreement within 5% of the experimental values was achieved.

The two-dimensional temperature field of the monolith is presented for a selected case in Fig. 5.9, representative of the continuum model simulations. Mean outlet gas temperatures computed for the individual reactor channels allowed for an estimate of the average monolith exhaust gas temperature. In the case of Fig. 5.9, the computed outlet gas temperature was at 1,089 K, as opposed to a measured value of 1,041 K for the same experimental conditions.

This figure also illustrates the strong variation of temperature in the radial direction over the axial length covering the rear half of the reactor. Flow field simulations of the single monolith channels for this case revealed that fuel is already consumed up to 99% at half the reactor length, which in turn renders the rest of the reactor volume a heat sink towards the environment. Reduced radial heat transfer (due to the spatially non-uniform thermal properties of the monolith) reduces the

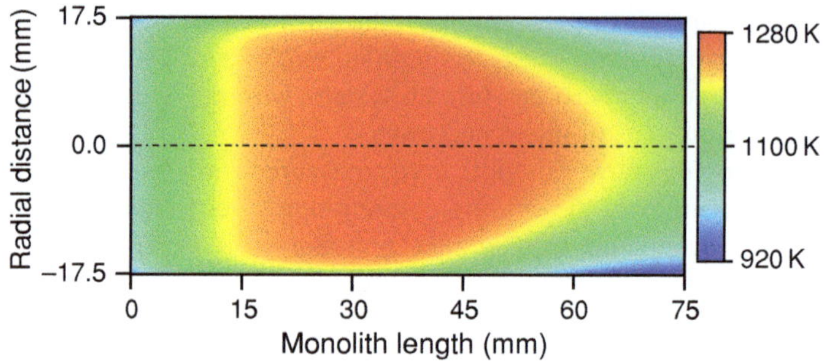

Fig. 5.9 Computed two-dimensional temperature distribution inside the catalytic monolith. The conditions are: $p_{IN} = 3.01$ bar, $\varphi = 0.20$, inlet temperature $T_{IN} = 713$ K, inlet velocity $U_{IN} = 3.21$ m/s, heat losses $h(T_w - T_\infty)$ with $h = 12$ W/m^{2}K and $T_\infty = 300$ K

amount of external heat losses; however these are large enough to drop the overall combustor efficiency down to 74% for this particular case. It should be noted that the large Lewis number of propane in its mixtures with air necessitates increased reactor lengths due to the reduced transverse transport of the fuel towards the reactor walls. The correct chemical expression derived from the first part of this work allows tuning the required reactor length with greater accuracy and thus reducing the unused part of the monolith, which serves only as a heat sink to the environment.

5.6 Conclusions

A propane-fueled, catalytic, mesoscale combustor was investigated numerically and experimentally to assess its applicability for a portable, gas-turbine-based power generation application. An optically accessible catalytic reactor was employed to establish a correct global step expression for propane combustion on platinum, for pressures up to 7 bar. Detailed parametric studies of a single catalytic channel revealed that, for a given reactor length, metallic materials are more favorable regarding maximum thermal power output and combustion efficiency. Radiation heat losses were more pronounced at the low inlet velocity regime and, moreover, had a greater impact at higher equivalence ratios. A subscale model of the catalytic combustor was constructed and tested under conditions dictated by the requirements of the power generation unit that met the required power output at the nominal mass throughput. Experiments revealed the positive effect of elevated operating pressures on the overall combustion efficiency, while significant heat losses of the monolith to the environment were evident. A continuum model was finally used to simulate the two-dimensional temperature field of the monolith, allowing for detailed description of the heat loss mechanisms. The latter tool, combined with the correct reaction rate expression for propane on platinum established earlier, provided the reactor length needed for full fuel conversion.

References

1. Schneider B, Karagiannidis S, Bruderer M, Dyntar D, Zwyssig C, Guangchun Q, Diener M, Boulouchos K, Abhari RS, Guzzella L, Kolar JW (2005) Ultra-high-energy-density converter for portable power, Power-MEMS 2005, Tokyo, Japan, 28–30 November 2005
2. Appel C, Mantzaras J, Schaeren R, Bombach R, Inauen A, Kaeppeli B, Hemmerling B, Stampanoni A (2002) An experimental and numerical investigation of homogeneous ignition in catalytically stabilized combustion of hydrogen/air mixtures over platinum. Combust Flame 128:340–368
3. Karagiannidis S, Mantzaras J, Jackson G, Boulouchos K (2007) Hetero-/homogeneous combustion and stability maps in methane-fueled catalytic microreactors. Proc Combust Inst 31:3309–3317
4. Siegel R, Howell JR (1981) Thermal radiation heat transfer. Hemisphere, New York, p 271
5. Reinke M, Mantzaras J, Schaeren R, Bombach R, Inauen A, Schenker S (2004) High-pressure catalytic combustion of methane over platinum: in situ experiments and detailed numerical predictions. Combust Flame 136:217–240
6. Kee RJ, Dixon-Lewis G, Warnatz J, Coltrin ME, Miller JA (1996) A fortran computer code package for the evaluation of gas-phase multicomponent transport properties, report no. SAND86-8246, Sandia National Laboratories
7. Groppi G, Tronconi E (1996) Continuous vs. discrete models on nonadiabatic monolith catalysts. AIChE 42:2382–2387
8. Coltrin ME, Kee RJ, Rupley FM (1996) Surface chemkin: a fortran package for analyzing heterogeneous chemical kinetics at the solid surface–gas phase interface, report no. SAND90-8003C, Sandia National Laboratories
9. Garetto TF, Rincon E, Apesteguia CR (2004) Deep oxidation of propane on Pt-supported catalysts: drastic turnover rate enhancement using zeolite supports. Appl Catal B Environ 48:167–174
10. Karagiannidis S, Mantzaras J, Bombach R, Schenker S, Boulouchos K (2009) Experimental and numerical investigation of the hetero-/homogeneous combustion of lean propane/air mixtures over platinum. Proc Combust Inst 32:1947–1955
11. Carroni R, Griffin T, Mantzaras J, Reinke M (2003) High-pressure experiments and modeling of methane/air catalytic combustion for power generation applications. Catal Today 83:157–170

Chapter 6
Hetero-/Homogeneous Combustion and Stability Maps in Methane-Fueled Catalytic Microreactors

6.1 Preface

In this chapter, a numerical investigation is undertaken to study the coupled catalytic and gas-phase combustion processes in a methane-fueled microreactor with catalytically active Pt walls. Simulations were carried out with the 2-D full elliptic model for both the gas and solid phases. Elementary hetero-/homogeneous chemical reaction schemes were included along with heat conduction in the walls, surface radiation heat transfer, and external heat losses. The main objectives were to investigate the interplay of hetero-/homogeneous combustion, transport, and heat transfer mechanisms in the microreactor and to delineate combustion stability maps in terms of the underlying parameters.

6.2 Numerical

The full elliptic 2-D laminar CFD code described in Chap. 3 was used to simulate the flow in a plane channel of length $L = 10$ mm, height $2b = 1$ mm and wall thickness $\delta = 0.1$ mm (see Fig. 6.1). A 2-D approach was also adopted for the heat conduction in the wall (solid thermal conductivity k_s). External heat losses were imposed at the outer wall surface, according to $h(T_w - T_\infty)$ with h a heat-transfer coefficient, T_w the local outer wall surface temperature and $T_\infty = 298$ K. Parametric studies were carried out by varying k_s, h, and the inlet velocity U_{IN}. Pressures of 1 and 5 bar were examined (the latter being of interest in turbine-based microreactors [1–3]) and the fuel-to-air equivalence ratio was $\varphi = 0.4$. The methane/air feed was preheated to 700 or 600 K (necessary to sustain combustion of fuel-lean CH_4/air mixtures over Pt [4]); in practice the preheat can be accomplished with heat recirculating microburners [5].

S. Karagiannidis, *Catalytic Microreactors for Portable Power Generation*,
Springer Theses, DOI: 10.1007/978-3-642-17668-5_6,

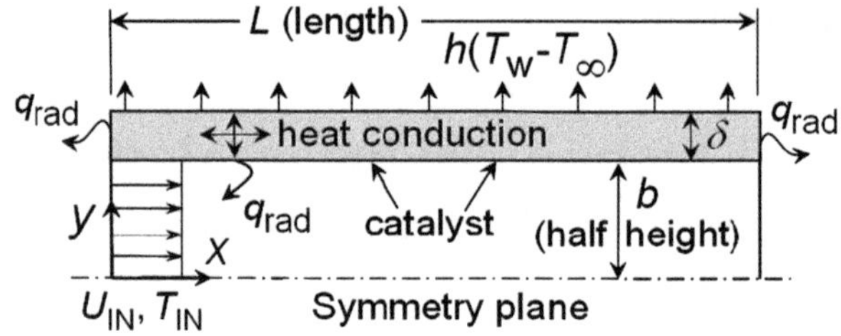

Fig. 6.1 Schematic of the microreactor configuration

Radiation heat transfer exchange between the discretized catalytic surface elements as well as between each surface element and the reactor inlet and outlet disk areas was accounted for by the net radiation method for diffuse-gray areas [6]. The inlet, outlet, and channel-element emissivities were all equal and the radiation exchange temperatures for the entry and outlet were equal to the corresponding mean gas temperatures.

In the bulk of the ensuing computations, the emissivity of all surfaces was constant at $\varepsilon = 0.6$. Gas radiative emission and absorption were not considered given the small optical paths and the large nitrogen dilution. Solid phase radiative boundary conditions were applied at the inlet and outlet vertical wall faces. Further details of the numerical model, along with the 2-D solid treatment have been provided elsewhere [7, 8].

The elementary heterogeneous scheme of Deutschmann et al. [9] (24 reactions, 11 surface and 9 gaseous species) was used to describe the oxidation of CH_4 on Pt. The C1/H/O elementary mechanism of Warnatz et al. [10] (26 species, 108 reactions) provided the platform for gas-phase chemistry. The aptness of this mechanism has been demonstrated for $p \geq 6$ bar and $0.05 \leq \varphi < 0.50$ [11]. To reproduce homogeneous ignition measurements at the lower pressure range 1 bar $\leq p <$ 6 bar, a correction in the kinetic parameters of the chain branching step $CHO + M \Leftrightarrow CO + H + M$ has been proposed [11], which was further supported by recent kinetic measurements [12]. The scheme of Warnatz with the aforementioned modification was valid at 1 bar $\leq p \leq$ 16 bar [11] and was thus used in the ensuing simulations. Gaseous and surface thermodynamic data were included in the provided schemes. Mixture-average diffusion, including thermal diffusion for the light species [13], provided the gas-phase transport model. Gas-phase and surface reaction rates were evaluated with CHEMKIN [14] and Surface-CHEMKIN [15], respectively.

An orthogonal staggered mesh of 24 $\times$ 84 points (in y and x, respectively) over half the gas-phase domain, with finer spacing towards the entry and the wall, was sufficient to produce a grid-independent solution. The computationally less-intensive solid heat conduction was solved on a 20 $\times$ 84 grid. Uniform inlet profiles were used for temperature, axial velocity and species mass fractions. No-slip was applied for both velocity components at the gas-wall interface ($y = b$) and zero-Neumann conditions were set at the outlet ($x = L$) and the plane of symmetry ($y = 0$). A combination of mainframe clusters and PCs (equivalent to twenty 2.6-GHz processors) was used and the CPU time per run was $\sim$10 h.

6.3 Results and Discussion

Combustion stability diagrams have been constructed by varying independently the external heat transfer coefficient, the inlet velocity, and the wall thermal conductivity. Figure 6.2 presents such diagrams for $k_s = 2$ W/mK and $\varepsilon = 0.6$. Therein, diagrams for $p = 1$ and 5 bar at $T_{IN} = 700$ K and also for $p = 5$ bar and $T_{IN} = 600$ K were computed by one-parameter continuation for a given U_{IN} until the critical extinction heat transfer coefficient was reached. To facilitate the forthcoming comparison of the $p = 1$ and 5 bar stability limits at a given mass throughput, two different velocity scales are employed in Fig. 6.2. The stability limits at large U_{IN} define blowouts due to insufficient residence times whereas at low U_{IN} limits are derived from heat-loss-induced extinction.

6.3.1 Coupling of Hetero-/Homogeneous Combustion and Heat Transfer Mechanisms

Streamwise profiles of the local catalytic and gas-phase methane conversion rates are provided in Fig. 6.3 for four combinations of h and U_{IN} (indicated by points A through D in Fig. 6.2). Cases A–C are at $p = 5$ bar and $T_{IN} = 700$ K, and Case D is at $p = 1$ bar and $T_{IN} = 700$ K. For direct comparisons with the catalytic surface conversion rates, the volumetric gaseous conversion rates have been integrated across the channel half-height in Fig. 6.3 to provide a conversion rate per unit reactor wall area. The corresponding catalyst surface temperatures are shown in Fig. 6.4. There is significant gas-phase methane conversion in Case A (Fig. 6.3A)

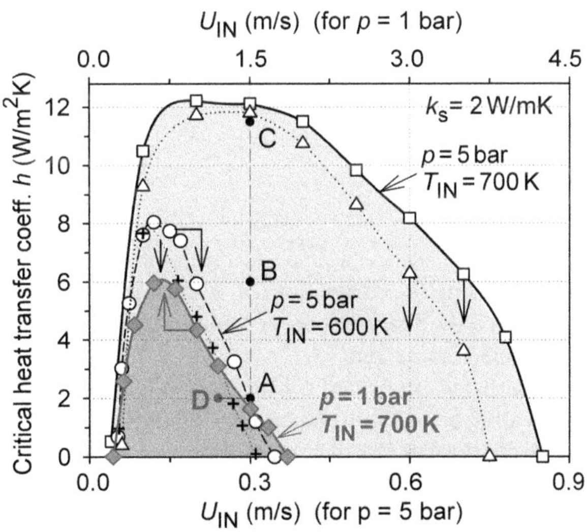

Fig. 6.2 Stability diagrams in terms of critical coefficient h and inlet velocity U_{IN}: $p = 5$ bar (*squares* $T_{IN} = 700$ K, *triangles* $T_{IN} = 700$ K without gas-phase chemistry, *circles* $T_{IN} = 600$ K, *crosses* $T_{IN} = 600$ K without gas-phase chemistry), $p = 1$ bar (*filled diamonds* $T_{IN} = 700$ K). The stable regimes for the 700 K cases are shown by the *shaded* areas

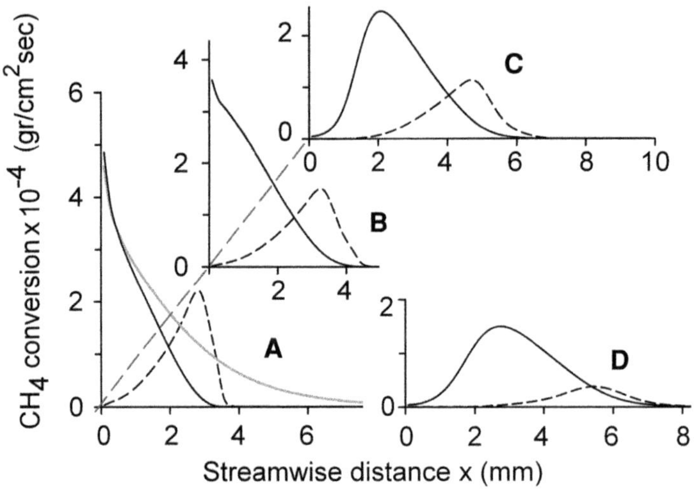

Fig. 6.3 Catalytic (*solid lines*) and gaseous (*dashed lines*) methane conversions for the four designated cases of Fig. 6.2: Cases *A*, *B*, *C* ($p = 5$ bar, $T_{IN} = 700$ K, $U_{IN} = 0.3$ m/s) and Case *D* ($p = 1$ bar, $T_{IN} = 700$ K, $U_{IN} = 1.2$ m/s). The *gray line* in (*A*) refers to pure catalytic combustion

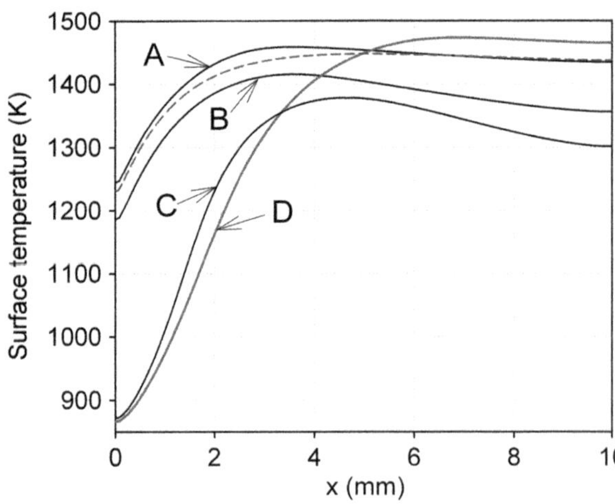

Fig. 6.4 Streamwise profiles of the catalyst surface temperature for Cases *A–D* of Fig. 6.2. The *dashed line* refers to Case *A* without gas-phase chemistry

that leads to the formation of a strong flame as illustrated in Fig. 6.5, which provides 2-D distributions of temperature, CH_4 and OH mass fractions, and gaseous heat release rate.

The flame, indicated by the heat release profile in Fig. 6.5d, is anchored near the wall at $x = 2$ mm, but due to catalytic depletion of methane in the vicinity of the wall (Fig. 6.5b), the most vigorous heat release occurs at the center of the channel farther downstream between $x = 3$ and 4 mm. All the same, the flame temperature at the channel center is only 16 K higher than the local wall

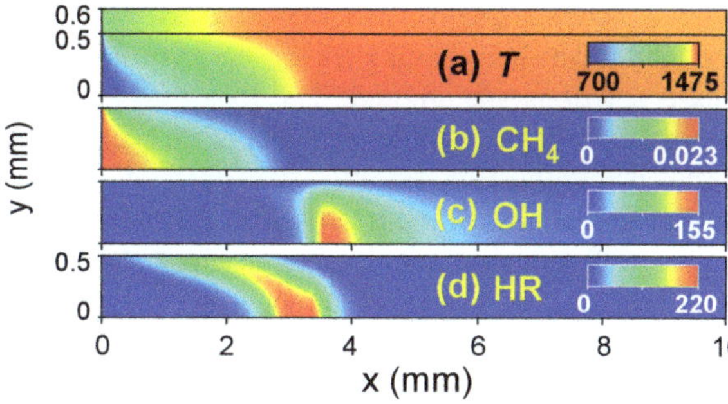

Fig. 6.5 Two-dimensional distributions of temperature (K), CH_4 and OH $\times$ 10^6 mass fractions, and gaseous heat release (W/cm^3) for Case *A* of Fig. 6.2. The channel-center and the gas-wall interface are at $y = 0$ and 0.5 mm, respectively. For clarity, the wall thickness in (*a*) is expanded by two

temperature (Fig. 6.5a). Axial conduction in the solid wall keeps the transverse temperature variation inside the wall to 5 K at $x = 0$ (at the channel exit the difference drops to 0.2 K). Therefore, the axial catalyst surface temperatures of Fig. 6.4 closely reflect the temperature inside the solid. Complete methane conversion (> 99.99%) is attained at the reactor exit for all Cases A–D, and the two reaction pathways compete for methane consumption. For the moderate velocity of Cases A–C ($U_{IN} = 0.3$ m/s), the relative contribution of the gaseous pathway decreases as the extinction limit is approached (Fig. 6.3) but it always remains significant and a flame is anchored inside the reactor just prior to extinction.

At higher U_{IN}, the hetero-/homogeneous chemistry coupling becomes more involved. Well-known aspects of this coupling include the promotion of gas-phase reactions due to the catalytically-induced exothermicity and the inhibition of gaseous reactions due to near-wall catalytic fuel depletion. In general, the hetero-/homogeneous radical coupling is weak in methane combustion over Pt [11, 16]. On the other hand, recent studies [11] have identified chemical coupling routes that are particularly relevant at microreactor conditions. The gas-phase combustion of methane can be roughly described by a two-step process, the incomplete oxidation of CH_4 to CO and the main heat-releasing oxidation of CO to CO_2 [17]. The catalytic pathway is a very efficient sink of homogeneously-produced CO due to the high sticking coefficient of this species (0.85 on Pt). By depriving CO from the gas phase, the catalyst inhibits the onset of homogeneous ignition. Nonetheless, the gas phase can still provide appreciable methane consumption without full flame formation via the incomplete oxidation of CH_4 to CO. This occurs (for both pressures) at higher U_{IN} (shorter residence times) over a range of heat transfer coefficients below the critical extinction value that increases with increasing U_{IN}. In Case D ($p = 1$ bar), for example, the ratio of the gaseous to the catalytic heat release is 0.10 while the corresponding methane consumption ratio is 0.21 (as illustrated by the heat release rates of the forthcoming Fig. 6.8 and the methane conversion rates of Fig. 6.3D) due to the incomplete gas-phase methane oxidation.

The gaseous pathway has been generally neglected in previous catalytic microreactor studies. However, as shown next, gas-phase reactions have significant impact on the microreactor performance under most of the conditions studied. When a flame is anchored in the reactor, the spatial extent of the heat release and methane consumption zones becomes narrower. Figure 6.3A (gray curve) shows the significant lengthening of the catalytic methane conversion profile when gas phase reactions are ignored. The localized gaseous heat release is more resilient against external heat losses than the more broadly distributed catalytic reactions. In terms of combustion stability, the presence of the gaseous pathway extends at moderate and low velocities the stability limits mildly. Moreover, the associated higher wall temperatures (as seen in Fig. 6.4 by the temperature profiles of Case A with and without gaseous reactions) are not endangering the reactor integrity. At higher velocities, the catalytic pathway may not attain complete methane conversion as the extinction limits are approached, due to finite-rate catalytic chemistry effects (although the mass-transport-limited catalytic conversion always guarantees complete methane consumption for the provided geometry and the velocities within the stable regimes of Fig. 6.2). In those higher velocity cases, the gaseous pathway has a more profound impact on extending the combustion stability limits by allowing for additional heat release. This is shown by two added stability diagrams in Fig. 6.2 ($p = 5$ bar, $T_{IN} = 700$ and 600 K), which were computed without gas-phase chemistry.

At $p = 5$ bar and $T_{IN} = 600$ K the stable combustion envelope narrows substantially, particularly in terms of U_{IN} (Fig. 6.2). It is hence evident that sufficient preheat of the feed is necessary for extended stability in fuel-lean CH_4/air microreactor combustion. The stability limits at $p = 1$ bar are much narrower than those at $p = 5$ bar ($T_{IN} = 700$ K), when the same mass throughput ($\rho_{IN}U_{IN}$) is considered. The reason is that both catalytic and gas-phase reactivities decline with decreasing pressure (the former scales as $\sim p^{+0.47}$ [16] and the latter as $\sim p^{+1.0}$ [17]). While Case A (5 bar) lies well-inside the combustion stability diagram of Fig. 6.2, the corresponding case at $p = 1$ bar ($h = 2$ W/m^{2}K) with the same mass throughput ($U_{IN} = 1.5$ m/s) is just outside the high-velocity extinction limit for the given h. Despite the reduced gas-phase reaction rates with lower pressure, the impact of the homogeneous pathway is still significant in defining the (narrower) stable combustion envelope.

6.3.2 *Influence of Solid Thermal Conductivity*

Stability limits are provided in Fig. 6.6 for two inlet velocities, $p = 5$ bar and $T_{IN} = 700$ K, in terms of the critical heat transfer coefficient for extinction. For low thermal conductivities ($k_s < 2$ W/mK), the reduced upstream heat transfer hinders catalytic ignition (light-off) and causes blowout. The stability limits at low k_s are narrower at higher inlet velocities (Fig. 6.6). In comparison to pure gas-phase combustion studies [18], there is a marked difference at the low k_s behavior, which is discussed qualitatively (since the aforementioned work refers to different geometry and operating conditions). In gas-phase combustion, the blowout limits extend over a narrower range of k_s ($\sim$0.4–0.8 W/mK) and are nearly independent of h (the blowout limit line is almost parallel to the h-axis). This is because low k_s

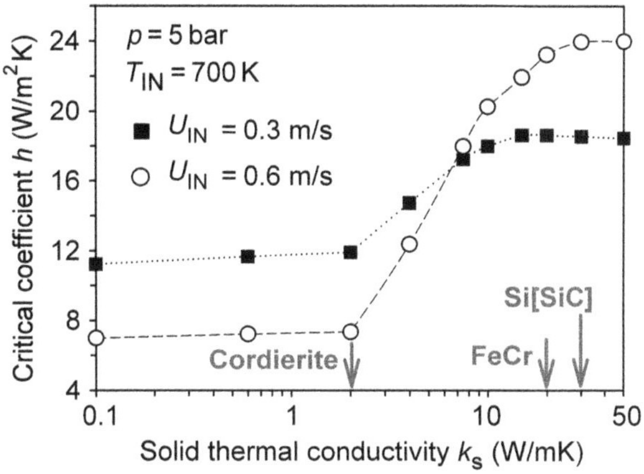

Fig. 6.6 Critical heat transfer coefficient versus solid thermal conductivity

reduce the conduction of heat away from the highly localized flame region and maintain moderate wall temperatures such that the external heat losses become less effective in removing heat from the walls. On the other hand, in the catalytic microreactor the distributed heat generation of the heterogeneous pathway provides a larger surface area with high temperatures, which in turn lead to enhanced external heat losses. Therefore, the external heat losses play a pivotal role in determining the stability of catalytic systems at low k_s. In addition, catalytic microreactors can achieve stable combustion at lower k_s compared to non-catalytic systems: the stable limits extend to $k_s < 0.1$ W/mK, but in practice there is little interest for such materials.

At higher k_s the extinction limits reach their larger extent for $k_s \sim 20$ W/mK and do not alter appreciably when k_s is increased to 50 W/mK. At low k_s the stability limits are defined by blowout and hence they are narrower at higher U_{IN} whereas at large k_s the limits are heat-loss-induced and hence are wider at higher U_{IN} (Fig. 6.6). In contrast, pure gas-phase combustion exhibits a peak at $k_s \sim 5$ W/mK [18] and a further increase to $k_s = 30$ W/mK narrows down considerably the stable regimes. The reason is that in pure gaseous combustion large k_s result in efficient removal of heat from the flame vicinity that eventually causes extinction. Despite the profound differences in the stability limit behavior of catalytic and non-catalytic microreactors, the maximum critical h ($\sim$24 W/m^{2}K, Fig. 6.6) is not very different from that reported in non-catalytic microreactors [18]. From the combustion stability point of view, materials such as FeCr alloy or Si(SiC) are well-suited for catalytic microreactors (Fig. 6.6).

6.3.3 Impact of Surface Radiation

Thermal radiation heat transfer from the hot catalytic surfaces is usually neglected in microreactor studies. However, it can have a strong impact not only on the stability limits but also on the combustion processes well-inside the stability envelope. In the foregoing analysis a surface emissivity $\varepsilon = 0.6$ was considered (its precise value depends on the type of washcoat, the catalyst loading, the surface treatment, etc.). In Case A, which is located well-below the stability limits, the catalyst surface temperatures are everywhere high (> 1,200 K, Fig. 6.4). Detailed energy balance of the solid (similar to that of the following Fig. 6.8) indicates that radiation is a net heat loss mechanism over the entire length of the catalytic surface because heat is radiated towards the significantly colder entry (700 K). Computed catalyst surface temperatures for Case A are shown in Fig. 6.7 for different emissivities.

The non-radiating ($\varepsilon = 0.0$) simulations indicate a substantially higher surface temperature (by 240 K at the entry and 110 K at the rear of the channel) compared to the $\varepsilon = 0.6$ results. The differences between the $\varepsilon = 0.6$ and 1.0 simulations are less pronounced but still significant (up to 100 K). The foregoing are in qualitative agreement with simulations of longer (aspect ratio of 28) catalytic monolithic channels [19]: therein, increasing ε from zero to unity caused large temperature drops at the entry (up to 500 K) and also moved the peak wall temperature farther downstream for those conditions where light-off was attained practically at $x \approx 0$ (as in Case A of Fig. 6.7, see also Fig. 6.3A). When the inlet and outlet sections were treated as black bodies ($\varepsilon_{IN} = \varepsilon_{OUT} = 1$, as in [19]) the curves of Case A (Fig. 6.7) differed by less than 50 K for $\varepsilon > 0.6$.

The role of radiation changes drastically as the stability limits are approached. Figure 6.8 provides (Case D) the energy balance terms for each solid phase slice—length Δx and thickness δ—surface heat generation $Q_{gen,cat}$, convection to the gas Q_{conv}, net radiation $Q_{rad,}$ heat loss Q_{loss} and integrated heat conduction Q_{cond} across the solid thickness δ. The gas-phase heat generation is not related to the solid energy balance and was included in Fig. 6.8 to facilitate the foregoing discussion of Sect. 6.3.1. Note that even close to the stability limit, the heat losses constitute a small (although decisive) part of the total solid energy balance.

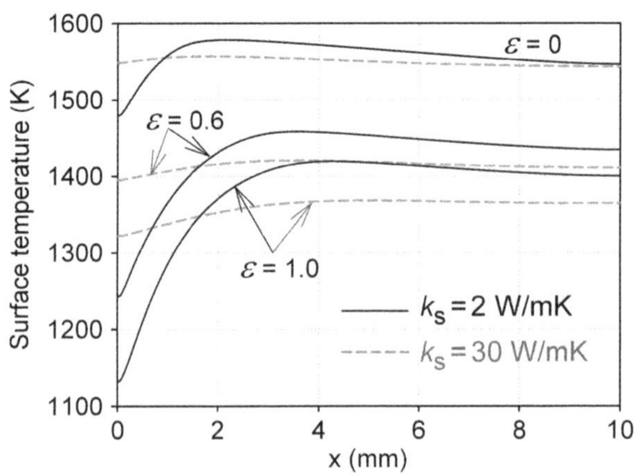

Fig. 6.7 Surface temperatures of Case *A* (*black solid lines*) for three emissivities. The *dashed gray lines* refer to $k_s = 30$ W/mK (all other parameters being the same with Case *A*)

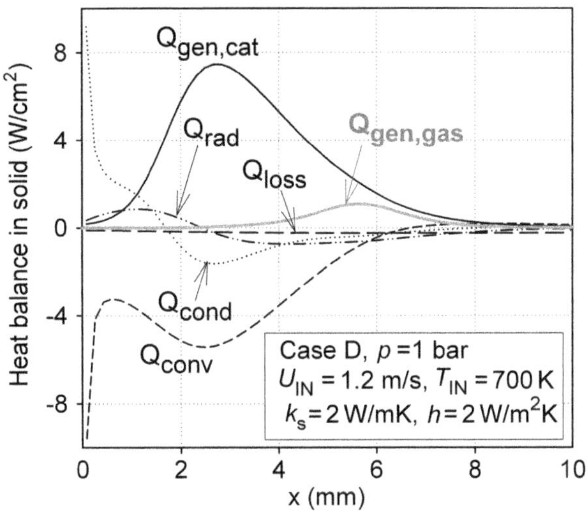

Fig. 6.8 Streamwise profiles of the solid energy balance for Case *D* of Fig. 6.2. The *y*-integrated gaseous heat release ($Q_{gen,gas}$) is additionally shown

In Case D the catalyst surface temperatures are low at the entry (< 900 K, Fig. 6.4) such that the radiation heat losses of the front surface elements to the inlet (700 K) are minimal. This leads to a positive net radiative heat flux for the first 2.5 mm and a negative for the remaining length (Fig. 6.8); heat is transferred from the rear of the channel to the front and this has a stabilizing effect on combustion. Computations with non-radiating surfaces for Case D ($\varepsilon = 0.0$) led to extinguishment of the combustion because upstream radiation heat transfer was necessary for catalyst light-off. The same behavior was also attested in Case C. Finally, at larger k_s the impact of radiation in reducing the entry temperatures becomes less severe (Fig. 6.7).

6.3.4 *Effect of Flow Confinement*

Additional simulations were carried out for channel half-heights of 0.30 and 0.15 mm. The main interest was to assess whether the homogeneous pathway still plays a role at such increased confinements; the detailed study of those confinements is outside the scope of this work. Computed 2-D distributions of the OH mass fraction and gaseous heat release rate are provided in Fig. 6.9 for those two confinements; the other operating parameters are the same as in Case A (Fig. 6.2). It is evident from Fig. 6.9 that gas-phase combustion can be sustained in catalytic microreactors at the sub-mm scale. The gas-phase combustion intensity decreases with increasing confinement (smaller channel height) as seen from the provided absolute OH and heat release magnitudes of Fig. 6.9. By increasing the confinement, the flames of Fig. 6.9 move upstream as a result of higher surface temperatures in the first 2 mm. This is an outcome of the decreased radiative heat

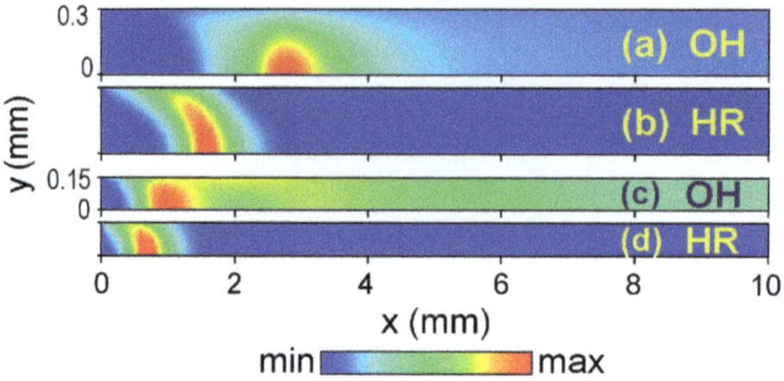

Fig. 6.9 Two-dimensional distributions of OH mass fraction and gaseous heat release (W/cm^3) for two half-heights: (*a*, *b*) 0.3 mm and (*c*, *d*) 0.15 mm. The other parameters are those of Case *A* (Fig. 6.2). The *color bar* indicates values from zero to (*a*) 5.2 × 10^{-5}, (*b*) 198, (*c*) 7.1 × 10^{-6}, and (*d*) 134

losses due to the reduction of the geometrical exchange view factors with the colder inlet (700 K) and the reduced convective transfer from the solid to the gas-phase due to lower mass throughput (constant U_{IN}).

6.4 Conclusions

The combustion and heat transfer processes in a catalytic microreactor (length and height of 10 and 1 mm, respectively) coated with Pt and fed with fuel-lean pre-heated CH_4/air mixtures were investigated numerically. A full-elliptic 2-D numerical model was employed that included detailed hetero-/homogeneous chemical reaction schemes and heat transfer mechanisms in the solid. Simulations at 1 and 5 bar have shown that the homogeneous reaction pathway could not be ignored at microreactor-relevant confinements and operating conditions. The presence of the gaseous pathway extended appreciably the blowout stability limits (high inlet velocity branch) and moderately the extinction limits (low inlet velocity branch). When considering the same mass throughput, the stable combustion envelope at 5 bar was substantially wider than at 1 bar due to the increased reactivities of both catalytic and gas-phase reaction pathways at elevated pressures. Stable combustion could be sustained with solid thermal heat conductivities (k_s) at least as low as 0.1 W/mK, while for k_s below 1.0 W/mK the allowable external heat losses became nearly independent of k_s. At k_s between 20 and 50 W/mK, a range that covers many metal and metallic/ceramic materials, the allowable heat losses reached their highest value. Comparison with literature pure gas-phase combustion studies has shown that the stability limits of catalytic microreactors are wider than those of similar non-catalytic devices. Surface radiation heat transfer has a dual role. For conditions far from extinction, surface radiation moderates the wall temperatures by providing a heat loss mechanism to the colder inlet. However, near the stability borders radiation can stabilize combustion by redistributing energy via transfer of heat from the hot rear to the colder front of the channel and aiding the catalyst light-off. Finally, with the coupling to the exothermic surface reactions, gas-phase combustion could be sustained even at sub-millimeter confinements of the catalytic microreactor.

References

1. Isomura K, Murayama M, Teramoto S, Hikichi K, Endo Y, Togo S, Tanaks S (2006) Experimental verification of the feasibility of a 100 W class micro-scale gas turbine at an impeller diameter of 10 mm. J Micromech Microeng 16:S254–S261
2. Peirs J, Waumans T, Vleugels P, Al-Bender F, Stevens T, Verstraete T, Stevens S, D'hulst R, Verstraete D, Fiorini P, Van Der Braembussche R, Driesen J, Puers R, Hendrick P, Baelmans M, Reynaerts D (2007) Micropower generation with microgasturbines: a challenge. J Mech Eng Sci 221:489–500
3. Schneider B, Karagiannidis S, Bruderer M, Dyntar D, Zwyssig C, Guangchun Q, Diener M, Boulouchos K, Abhari RS, Guzzella L, Kolar JW (2005) Ultra-high-energy-density converter for portable power. In: Power-MEMS 2005, 28–30 November, Tokyo
4. Deutschmann O, Schmidt R, Behrendt F, Warnatz J (1996) Numerical modeling of catalytic ignition. Proc Combust Inst 26:1747–1754
5. Ahn JM, Eastwood C, Sitzki L, Ronney PD (2005) Gas-phase and catalytic combustion in heat-recirculating burners. Proc Combust Inst 30:2463–2472
6. Siegel R, Howell JR (1981) Thermal radiation heat transfer. Hemisphere, New York, p 271
7. Dogwiler U, Benz P, Mantzaras J (1999) Two-dimensional modeling for catalytically stabilized combustion of a lean methane-air mixture with elementary homogeneous and heterogeneous chemical reactions. Combust Flame 116:243–258
8. Schneider A, Mantzaras J, Jansohn P (2006) Experimental and numerical investigation of the catalytic partial oxidation of CH_4/O_2 mixtures diluted with H_2O and CO_2 in a short contact time reactor. Chem Eng Sci 61:4634–4646
9. Deutschmann O, Maier LI, Riedel U, Stroemman AH, Dibble RW (2000) Hydrogen assisted catalytic combustion of methane on platinum. Catal Today 59:141–150
10. Warnatz J, Dibble RW, Maas U (1996) Combustion, physical and chemical fundamentals, modeling and simulation. Springer, New York
11. Reinke M, Mantzaras J, Bombach R, Schenker S, Inauen A (2005) Gas-phase chemistry in catalytic combustion of methane/air mixtures over platinum at pressures of 1 bar to 16 bar. Combust Flame 141:448–468
12. Friedrichs G, Herbon JT, Davidson DF, Hanson RK (2002) Quantitative detection of HCO behind shock waves: the thermal decomposition of HCO. Phys Chem Chem Phys 4:5778–5788
13. Kee RJ, Dixon-Lewis G, Warnatz J, Coltrin ME, Miller JA (1996) A Fortran computer code package for the evaluation of gas-phase multicomponent transport properties. Report No. SAND86-8246; Sandia National Laboratories
14. Kee RJ, Rupley FM, Miller JA (1996) Chemkin II: a Fortran chemical kinetics package for the analysis of gas-phase chemical kinetics. Report No. SAND89-8009B; Sandia National Laboratories
15. Coltrin ME, Kee RJ, Rupley FM (1996) Surface chemkin: a Fortran package for analyzing heterogeneous chemical kinetics at the solid surface-gas phase interface. Report No. SAND90-8003C; Sandia National Laboratories
16. Reinke M, Mantzaras J, Schaeren R, Bombach R, Inauen A, Schenker S (2004) High-pressure catalytic combustion of methane over platinum: in situ experiments and detailed numerical predictions. Combust Flame 136:217–240

17. Westbrook CK, Dryer FL (1981) Simplified reaction mechanisms for the oxidation of hydrocarbon fuels in flames. Combust Sci Technol 27:31–43
18. Norton DG, Vlachos DG (2003) Combustion characteristics and flame stability at the microscale: a CFD study of premixed methane/air mixtures. Chem Eng Sci 58:4871–4882
19. Boehman AL (1998) Radiation heat transfer in catalytic monoliths. AIChE 44:2745–2755

Chapter 7
Stability of Hetero-/Homogeneous Combustion in Propane- and Methane-Fueled Catalytic Microreactors: Channel Confinement and Molecular Transport Effects

7.1 Preface

A detailed parametric study is undertaken in this chapter, using the full elliptic 2-D CFD code for both gas-phase and solid domains, in order to delineate the stable combustion regimes of propane-fueled catalytic microreactors at pressures 1 and 5 bar (pressures up to 5 bar are of interest to recuperated microturbine systems [1–3]), channel confinements 1.0 and 0.3 mm and wall thermal conductivities 2 and 16 W/mK. Methane simulations are also included, so as to exemplify the significant differences in both chemical and transport properties on microcombustor stability. The main objectives are to assess the effect of high pressure operation, molecular transport and gas-phase chemistry on the stability of propane-fueled catalytic microreactors and to study the impact of increased geometrical confinement and high wall thermal conductivity on the non-adiabatic reactor operation. Particular objectives were to quantify the differences between the two fuels in terms of reactor stability and performance.

7.2 Numerical Model

A full-elliptic, 2-D steady laminarl CFD code [4–6] was employed to simulate the flow domain in a plane channel configuration having a total length $L = 10$ mm, channel heights of either $2b = 0.3$ or 1 mm, and wall thickness of $\delta = 0.1$ mm (Fig. 7.1). The first 1 mm was catalytically inert and the remaining $L_a = 9$ mm was coated with platinum. Due to symmetry, only half of the flow domain was solved. The flow solver was coupled to a 2-D heat conduction solver for the solid substrate (wall thermal conductivity k_s). External heat losses were applied to the outer channel surface as heat flux $h(T_w - T_\infty)$ using an effective heat transfer coefficient h, with T_w the outer surface temperature and T_∞ set to 298 K. Propane and methane fuels

S. Karagiannidis, *Catalytic Microreactors for Portable Power Generation*, 67
Springer Theses, DOI: 10.1007/978-3-642-17668-5_7,

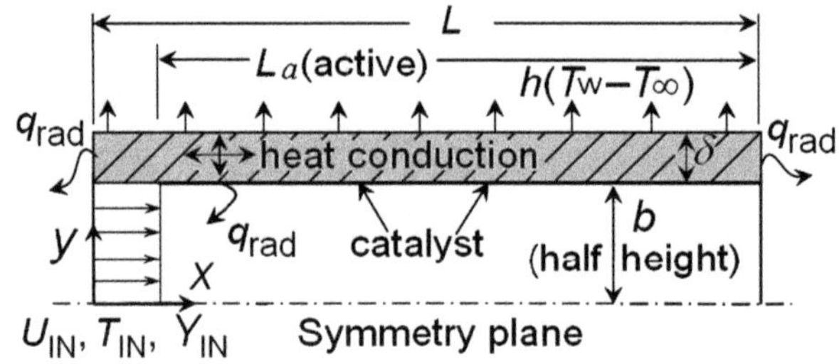

Fig. 7.1 Schematic of the channel-flow microreactor

were examined (the former of prime interest for portable power generation), with a fuel-to-air equivalence ratio $\varphi = 0.4$, preheat $T_{IN} = 700$ K and pressures $p = 1$ and 5 bar. Solid thermal conductivities $k_s = 2$ and 16 W/mK were considered, corresponding to cordierite (ceramic) and FeCr alloy (metallic) walls, respectively. The inlet velocity U_{IN}, external heat transfer coefficient h and fuel type were the independent variables of the parametric study, under various microreactor configurations of channel confinement and wall thermal conductivity. A summary of the reactor configurations studied is provided in Table 7.1. The net radiation method for diffuse-gray areas [7] accounted for radiation exchange between the discretized channel wall elements themselves and also between the elements and the inlet and outlet channel enclosures. Emissivities for the enclosures were set equal to the emissivity of the channel wall surfaces, $\varepsilon_{IN} = \varepsilon_{OUT} = \varepsilon = 0.6$. Details on the radiation model have been in Sect. 3.1.

Catalytic combustion of lean propane/air mixtures over platinum is modeled using a recently proposed global step reaction rate expression [8], validated over 1 bar $\leq p \leq$ 7 bar. Major hetero-/homogeneous chemistry interactions arise from the near-wall depletion of fuel due to surface reactions (an effect hindering gas-phase ignition) and the presence of catalytically produced major combustion species CO_2 and H_2O [9], with the radical coupling between the two chemical pathways being particularly weak [10, 11]. This level of chemical interaction allowed for the successful coupling of the single catalytic step with a detailed gas-phase reaction scheme, as shown by model comparisons with in situ spatially resolved measurements of major and minor species [8]. For propane, the detailed C_3 gas-phase chemical scheme by Qin et al. [12] was employed, comprising of 70 species in 14 irreversible and 449 reversible elementary reactions. For methane, the elementary heterogeneous scheme of Deutschmann et al. [13] was used (24 reactions, 11 surface and 9 gaseous species) for lean methane/air combustion over platinum, coupled to the C_1 elementary gas-phase mechanism of Warnatz et al. [14] (26 species, 108 reactions). The former mechanism was validated in [15] while the latter has been adapted with corrections for $p \leq 6$ bar, further supported by recent kinetic measurements, so as to reproduce homogeneous ignition measurements [16]. Mixture-average diffusion, including thermal diffusion for light species [17] was used, while surface and gas-phase reaction rates were evaluated using Surface-CHEMKIN [18] and CHEMKIN [19], respectively.

Table 7.1 Catalytic microreactor configuration

Length L (mm)	10
Active length L_a (mm)	9
Wall thickness δ (mm)	0.1
Channel half-height b (mm)	0.5, 0.3
Solid thermal conductivity k_s (W/mK)	2 (cordierite), 16 (FeCr alloy)
Inlet pressure p (bar)	1, 5
Inlet temperature T_{IN} (K)	700
Channel surface emissivity ε	0.6
Fuel/air equivalence ratio φ	0.4
Fuel	C_3H_8, CH_4

An orthogonal staggered grid with 100×24 points in x- and y-direction, respectively, over the channel half-height was sufficient to produce a grid-independent solution. Finer spacing towards the channel wall and the entry section was used. The temperature field of the 10×0.1 mm^2 solid was solved on a 100×20 grid. At the inlet, uniform profiles of species, temperature and axial velocity were applied, while zero-Neumann conditions were set at the outlet ($x = L$) and plane of symmetry ($y = 0$). No-slip was applied for both velocity components at the gas-wall interface ($y = b$). All computations were performed on a cluster with 16 Xeon CPUs at 3.0 GHz, with the computational time ranging from 10 up to 32 h on a single processor for methane and propane fuels, respectively.

7.3 Results and Discussion

The stable combustion regimes of propane- and methane-fueled catalytic microreactors were delineated for pressures of 1 and 5 bar under reactor configurations (b, k_s) and operating conditions relevant to power generation microdevices [3]. Combustion stability diagrams of inlet velocity U_{IN} versus external heat loss coefficient h were computed for propane- and methane-fueled catalytic microreactors via one-parameter continuation of these parameters until they assumed values where combustion was either extinguished or blown out of the numerical domain.

Two-dimensional distributions of fuel and OH mass fractions are shown along with the gas-phase temperature for propane- and methane-fueled catalytic microreactors in Fig. 7.2a, b, respectively, for a channel height of 0.3 mm. Due to heat conduction in the 0.1 mm thin solid wall and for the present steady calculations, transverse temperature differences inside the channel wall were in all cases ≤ 4 K.

Corresponding streamwise profiles of the local catalytic and gas-phase (the latter integrated over the channel half-height b) fuel conversion rates for the conditions of Fig. 7.2 are provided in Fig. 7.3. Catalytic reactions are initiated after the first inert

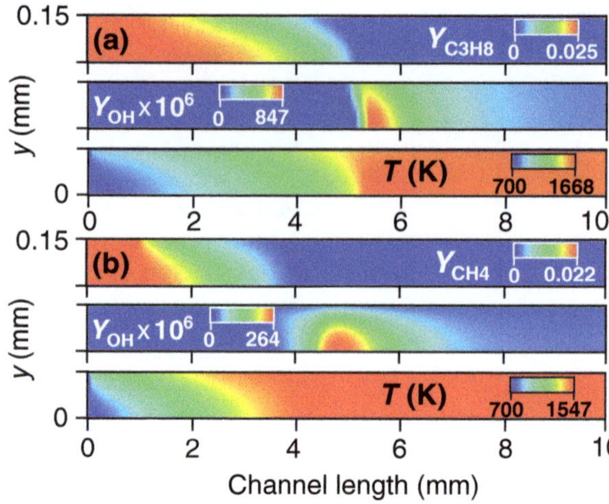

Fig. 7.2 Two-dimensional distribution of fuel and OH $\times$ 10^6 mass fractions and gas-phase temperature for (a) propane- and (b) methane-fueled microreactors. Gas-wall interface: $y = 0.15$ mm, axis of symmetry: $y = 0$ mm. Other conditions: $p = 5$ bar, $U_{IN} = 2.3$ m/s, $h = 4$ W/m^{2}K, $k_s = 16$ W/mK

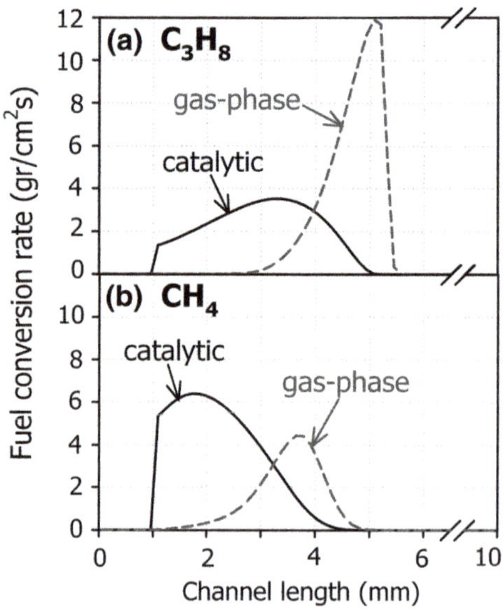

Fig. 7.3 Catalytic (*solid lines*) and gas-phase (*dashed lines*) conversion rates for (a) C_3H_8 and (b) CH_4. Conditions same as in Fig. 7.2. Flame positions at: $x = 5.1$ (C_3H_8), 3.9 mm (CH_4)

1 mm length. Despite the large channel confinement ($b = 0.15$ mm), the catalytic exothermicity heats up the walls and allows strong flames to be established at $x = 5.1$ and 3.9 mm for C_3H_8 and CH_4, respectively, evident by the high OH radical concentrations and the peak in the homogeneous fuel conversion rates at these positions (Fig. 7.3). Complete fuel conversion is attained in both cases. Figures 7.2 and 7.3 indicate a qualitative difference between methane and propane hetero-/homogeneous combustion. Catalytic reactions (peaking at mass-transport-limited conditions) convert most of methane, while for propane a much stronger flame is established (Fig. 7.3) due to the lower transport rate of propane to the catalytic surface compared

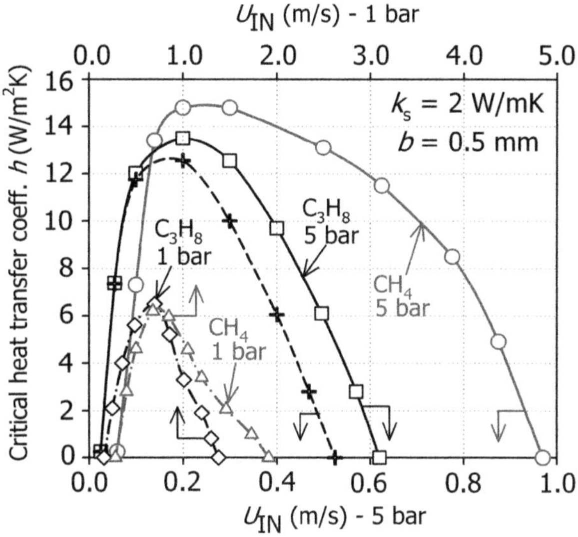

Fig. 7.4 Stability maps in terms of inlet velocity U_{IN} versus critical transfer heat coefficient h for 1 and 5 bar, channel confinement $b = 0.5$ mm and wall thermal conductivity $k_s = 2$ W/mK. Propane: *squares*: 5 bar, *diamonds*: 1 bar. Methane: *circles*: 5 bar, *triangles*: 1 bar. *Crosses/dashed line*: C_3H_8 stability curve at 5 bar without gas-phase chemistry. Constant $\rho_{IN}U_{IN}b$

to methane, resulting in more fuel being available for gas-phase combustion; the implications of this behavior are discussed next.

7.3.1 Stability Maps: Pressure, Gas-Phase Chemistry and Fuel Transport Properties Effects

Stability diagrams for pressures $p = 1$ and 5 bar are presented in Fig. 7.4, for C_3H_8- and CH_4-fueled microreactors, where the critical heat transfer coefficient h is plotted against the inlet velocity U_{IN}. Wall thermal conductivity is set to $k_s = 2$ W/mK (corresponding to cordierite ceramic), with a channel half-height $b = 0.5$ mm. In order to facilitate direct comparisons between different pressures, two scales of inlet velocities are provided in this figure (one for each pressure), such that the mass throughput ($\rho_{IN}U_{IN}b$) is constant along the x-axis.

In a fashion similar to earlier studies for methane [4], Fig. 7.4 indicates that by increasing the operating pressure of the propane-fueled microreactor from 1 bar (diamonds) to 5 bar (squares) a significant, almost two-fold increase on the maximum allowable external heat transfer coefficient h is achieved for the same mass throughput. At the same time, the upper mass inflow limit before blowout is extended substantially ($\sim$124%). Recent kinetic studies revealed a positive $p^{+0.75}$ dependence of propane reactivity on platinum [8], while earlier studies suggested a corresponding positive $p^{+1.75}$ dependence for gas-phase reactions [20]. The enhanced microreactor stability at modest pressures can thus be attributed to the positive pressure dependence of the hetero-/homogeneous propane reactivity.

The stable combustion regime at $p = 1$ bar is essentially a subset of that at $p = 5$ bar, and the same holds true for methane.

Until recently, gas-phase chemistry was generally not considered in catalytic microreactor simulations, even though experiments provided evidence towards the presence of vigorous homogeneous reactivity under certain operating conditions [21]. Gas-phase chemistry was shown to significantly extend the blowout limits of methane-fueled catalytic microreactors [4], and to a lesser extent the extinction limits at the low inlet velocities. The stability curve at 5 bar has been computed anew for the case of propane, but with homogeneous chemistry turned off (Fig. 7.4, crosses). The stable combustion regime is evidently narrower when gas-phase chemistry is neglected, following qualitatively the same trend as in the case of methane. Flames anchored inside the channel provide additional heat release at high inlet velocities (the associated low residence times could allow for fuel breakthrough when only surface reactions are considered), but most importantly flames provide a localized heat release zone which is shielded by heat losses to the surroundings, since the gas thermal conductivity is appreciably lower from that of the solid walls. Quantitatively, the stability of propane-fueled catalytic microreactors benefits significantly more from the presence of gas-phase reactions than their methane-fueled counterparts: while earlier methane studies [4] showed that gas-phase chemistry increases the critical heat transfer coefficient h by $\sim$50% in the blowout branch of the stability curve at the maximum U_{IN} limit ($h = 0$), for propane the corresponding increase is $\sim$65%. Compared to methane, propane has higher gas-phase reactivity [20], allowing vigorous homogeneous combustion at relatively low reactor temperatures combined with the reduced conversion of propane on the catalytic surface (due to its lower molecular diffusivity as will be shown later, see Fig. 7.3) which leaves more fuel available for gas-phase reactions.

The most profound observation in Fig. 7.4 is the significant difference between propane and methane stability limits. At the extinction branch of the stability limits ($U_{IN} < 0.2$ m/s at 5 bar), the propane-fueled catalytic reactor is substantially more robust against external heat losses. Despite the higher catalytic [22] and gas-phase reactivity of propane when compared to methane, at the blowout branch this trend is reversed. Methane has much broader limits, in terms of maximum allowable critical heat transfer coefficient h and of inlet velocity U_{IN}. This behavior of catalytic microreactors comes in stark contrast to pure gas-phase microreactors [23], whereby the stability diagrams of methane are substantially narrower than those of propane.

To understand the aforementioned behavior, additional calculations were performed with the Surface Perfectly Stirred Reactor (SPSR) package of CHEMKIN [24], using the coupled hetero-/homogeneous reaction schemes for both fuels. Results are presented in Fig. 7.5, with input conditions for all calculations $p = 5$ bar and $\varphi = 0.4$. Two surface-to-volume ratios were studied, $S/V = 20$ and 66.7 cm^{-1}, corresponding to plane channels with confinements of $b = 0.5$ and 0.15 mm, respectively. For a fixed reactor temperature, residence times $\tau_r = 5$ and 20 ms were chosen, corresponding to intermediate and high reactor inlet velocities. Constant temperatures in the SPSR were varied between $T_r = 850$

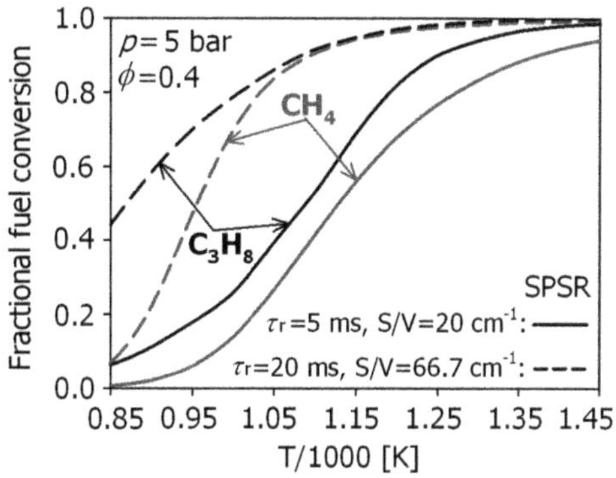

Fig. 7.5 Surface perfectly stirred reactor (SPSR) calculations of fractional fuel conversion for C_3H_8/ and CH_4/air mixtures at $p = 5$ bar, $\varphi = 0.4$, two surface-to-volume ratios (*S/V*) and two residence times (τ_r). *Solid lines*: $\tau_r = 5$ ms, $S/V = 20\ cm^{-1}$. *Dashed lines*: $\tau_r = 20$ ms, $S/V = 66.7\ cm^{-1}$

and 1,450 K. It is evident from Fig. 7.5 that, for all configurations, propane conversion assumes greater values thanks to its higher heterogeneous and homogeneous reactivity, with methane conversion approaching that of propane only for $T_r \geq 1{,}050$ K and $\tau_r = 20$ ms. It can thus be concluded that, while chemistry can explain the greater combustion stability of propane at low U_{IN} (and hence high τ_r), it ceases to be the controlling factor at shorter τ_r.

Recent studies on propane-fueled catalytic microcombustors [25] at atmospheric pressure hinted towards the importance of molecular transport effects in reactors of sub-millimeter scales. Since in lean methane/air combustion the fuel has Lewis number $Le_{CH_4} \approx 0.97$, the methane transverse transport towards the catalytic channel surface will be considerably higher than that of propane, which is highly diffusionally imbalanced with $Le_{C_3H_8} \approx 1.82$. It can thus be expected that when the residence time in the channel has the same order of magnitude as the transverse fuel diffusion times, fuel conversion and combustion stability will be impacted primarily by the transport rather than the chemical properties of the fuel.

In order to assess the effect of fuel (being the deficient reactant in lean combustion) diffusivity, additional simulations were performed for a channel with $k_s = 2$ W/mK and $b = 0.5$ mm, at $p = 5$ bar, adiabatic outer channel surface ($h = 0$) and with gas-phase reactions turned off. Three fuels were examined: C_3H_8, CH_4 and an artificial fuel $C_3H_8(*)$, the latter having the same chemical and thermodynamic properties as C_3H_8 but with a Lewis number of unity by setting its mass diffusivity equal to the mixture thermal diffusivity over the entire gas-phase computational domain. In Fig. 7.6, the fractional fuel conversion is plotted versus inlet velocity, along with the streamwise wall temperature profiles at the solid-gas interface for the three fuels studied at $U_{IN} = 0.5$ m/s, a value close to the maximum allowable for C_3H_8. Although the methane-fueled catalytic microreactor can only sustain combustion for a total fuel conversion $\geq$ 82%, blowout occurs for inlet velocities $\geq$ 0.945 m/s, a value 80% higher than the maximum inlet velocity the propane-fueled microreactor can sustain before blowout. Wall temperature profiles in the inset of Fig. 7.6 show significantly lower wall temperatures along the entire channel length for C_3H_8 compared to CH_4 and $C_3H_8(*)$, at an inlet velocity close to the blowout limit.

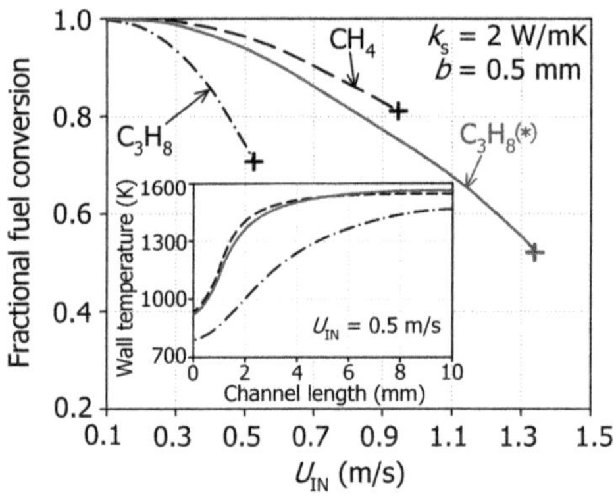

Fig. 7.6 Fractional fuel conversion versus U_{IN}, for a microreactor with k_s = 2 W/mK, b = 0.5 mm and h = 0 W/m^{2}K at p = 5 bar with catalytic reactions only. Crosses denote maximum U_{IN} sustainable for each fuel. *Dash-dotted line*: C_3H_8. *Dashed line*: CH_4. *Solid line*: C_3H_8(*) (Le = 1). *Inset*: wall temperature profiles at U_{IN} = 0.5 m/s for C_3H_8, CH_4 and C_3H_8(*)

For the lean catalytic combustion of diffusionally imbalanced fuels such as propane, with Le > 1, theory dictates surface temperatures at the reactor entry section lower than the adiabatic flame temperature even at transport-limited operation [9, 26, 27] (neglecting heat conduction in the solid substrate). This behavior, when considering the added effects of finite-rate chemistry and radiation losses towards the cold inlet enclosure (at T_{IN} = 700 K), results in wall temperatures low enough that push the catalytic reaction zone farther downstream and thus render the reactor prone to blowout at higher U_{IN}. On the other hand, the higher diffusivity of methane maintains high front-end surface temperatures (Fig. 7.6). In the case of the artificial fuel C_3H_8(*) with $Le_{C_3H_8(*)} = 1$, the combination of high catalytic reactivity with transport properties close to that of methane result in an almost threefold increase of the maximum U_{IN} for C_3H_8(*) compared to C_3H_8 and almost 40% increase compared to CH_4 (Fig. 7.6). The wall temperature profiles for C_3H_8(*) closely match those of methane (inset of Fig. 7.6), clearly showing that the diffusive transport properties of the fuel can have a greater impact on microreactor stability compared to chemical properties. It is stressed that the transport effect is much more pronounced in microreactors due to their particularly short length; in other industrial applications (e.g., large-scale power generation) where lengths of 100 mm or more are typical this effect diminishes.

By comparing the stability diagrams of C_3H_8 and CH_4 at 1 and 5 bar in Fig. 7.4, it is evident that differences are more pronounced as the operating pressure increases. While propane remains more resilient towards external heat losses at low U_{IN}, an increase in reactor pressure is substantially more beneficial for methane- than propane-fueled microreactors. Transverse profiles of CH_4 and C_3H_8 mass fractions are presented in Fig. 7.7 for U_{IN} = 1.2 and 0.5 m/s at p = 1 and 5 bar, respectively; the selected inlet velocities are close to the blowout limits of propane for the two pressures (see Fig. 7.4).

Profiles are presented for three streamwise positions, for which the local gas-phase conversion rate does not account for more than 5% of the corresponding catalytic one. As pressure increases to 5 bar, the reactor tends towards transport-controlled fuel conversion owing to increased catalytic reactivity for both fuels,

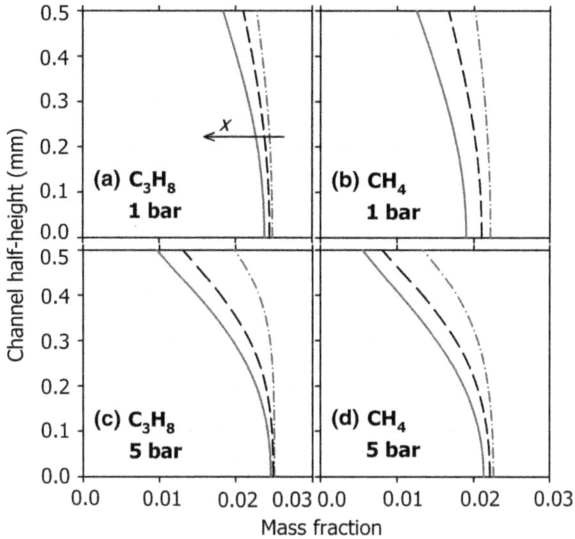

Fig. 7.7 C_3H_8 and CH_4 transverse mass fraction profiles at three streamwise positions, for a reactor with $b = 0.5$ mm, $k_s = 2$ W/mK and $h = 0$ W/m^{2}K. (a, b) $U_{IN} = 1.2$ m/s, $p = 1$ bar, $x = 1.98$ (*dash-dotted*), 2.50 (*dashed*), 3.01 mm (*solid lines*). (c, d) $U_{IN} = 0.5$ m/s, $p = 5$ bar, $x = 1.09$ (*dash-dotted*), 1.60 (*dashed*) and 1.98 mm (*solid lines*)

manifested by the reduced amounts of fuel at the gas-wall interface ($y = 0.5$ mm in Fig. 7.7c, d); conversely, fuel conversion at 1 bar is essentially kinetically controlled for both fuels (Fig. 7.7a, b). This shifting towards transport-limited fuel conversion at higher pressures enhances the impact of transport properties on the combustion stability, thus allowing methane-fueled microreactors to benefit more from the increased surface reactivity than the propane-fueled ones.

7.3.2 *Stability Maps: Impact of Channel Confinement and Wall Thermal Conductivity*

Additional computations were performed for microreactors with larger channel confinement ($b = 0.15$ mm), for two wall thermal conductivities ($k_s = 2$ and 16 W/mK). The stable combustion regimes of C_3H_8- and CH_4-fueled catalytic microreactors at $p = 5$ bar for the above mentioned configurations are presented in Fig. 7.8. Stability curves constructed with homogeneous chemistry turned off show that, even in a narrow channel with $b = 0.15$ mm, gaseous reactions can extend the blowout limits of propane catalytic microreactors (triangles/dashed and triangles/solid curves in Fig. 7.8), albeit with a reduced impact compared to reactors with smaller confinements (see Fig. 7.4).

To facilitate comparison with the nominal case studied, the stability curve of a propane-fueled microreactor with $b = 0.5$ mm, $k_s = 2$ W/mK and $p = 5$ bar is also provided in Fig. 7.8; the two velocity scales are such that any point in the horizontal axis has the same mass throughput either for $b = 0.15$ mm or for

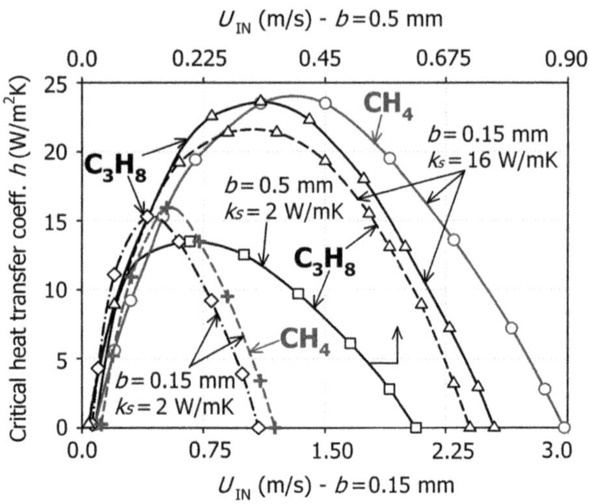

Fig. 7.8 Combustion stability diagrams in terms of inlet velocity U_{IN} versus critical heat transfer coefficient h for C_3H_8 and CH_4 fuels at $p = 5$ bar, $b = 0.15$ mm, and $k_s = 2$ and 16 W/mK. C_3H_8: *triangles* (16 W/mK) and *diamonds* (2 W/mK). CH_4: *circles* (16 W/mK) and *crosses* (2 W/mK). *Dashed lines/triangles*: C_3H_8 stability curve ($k_s = 16$ W/mK) without gas-phase reactions. *Squares*: C_3H_8 stability curve for $k_s = 2$ W/mK, $b = 0.50$ mm. (provided for comparison from Fig. 7.4)

0.50 mm. Qualitatively the same trend is retained at the higher channel confinements: for sufficiently high residence times (low U_{IN}) propane tolerates higher external heat losses, while at higher inlet velocities methane benefits from increased transverse transport towards the catalytic surface, thus having a wider combustion stability envelope in terms of maximum U_{IN} and critical h. On the other hand, the differences between C_3H_8- and CH_4-fueled reactors are less pronounced than for $b = 0.5$ mm in Fig. 7.4, for both values of k_s studied. Since the characteristic transverse diffusion time scales as b^2/D_F, where D_F the diffusivity of propane or methane, it is expected that as the channel half-height decreases, molecular transport-induced phenomena will become less important, and this is clearly evident in Fig. 7.8.

A comparison of stability limit curves between $b = 0.5$ and 0.15 mm in Fig. 7.8 reveals that, for the same mass throughput, reducing the channel half-height enhances combustion stability towards external heat losses at the extinction branch, allowing for a higher value of the critical coefficient h. The increased transverse diffusion transport of the fuel on the reactor walls at $b = 0.15$ mm, with the associated increased heat release, compensates more efficiently for the heat lost from the outer channel surface. At the blowout branch, however, the residence times of the fuel/air mixture become too short and blowout occurs for mass throughputs lower than for a reactor with $b = 0.5$ mm, a result valid for both for C_3H_8 and CH_4 (see Figs. 7.4 and 7.8).

Increasing the wall thermal conductivity from 2 to 16 W/mK greatly broadens the combustion stability envelope, for reactors with $b = 0.15$ mm. Both propane and methane reactors benefit from the increased upstream heat transfer through the reactor wall and the associated enhanced preheating of the incoming fuel/air mixture. The maximum allowable U_{IN} is increased by almost threefold, while the maximum critical h also increases by ~60% for both fuels. However, at low U_{IN} the extinction limits of reactors with high k_s are lower than their low k_s counterparts.

In Fig. 7.9, channel wall temperature profiles are presented for cordierite and FeCr alloy microreactors with $b = 0.15$ mm and $h = 8$ W/m^{2}K. For each fuel considered, two inlet velocities U_{IN} are presented, $U_{IN} = 0.3$ and 0.9 m/s for CH_4 and $U_{IN} = 0.2$ and 0.8 m/s for C_3H_8; for the selected h value, the lower inlet velocities are near the extinction limit of high k_s reactors while the higher ones close to the blowout limit of low k_s reactors. At low U_{IN}, low thermal conductivity microreactors benefit from a localized reaction zone and the associated high surface temperatures, which in turn promote stable combustion even under elevated external heat losses (see, e.g., $U_{IN} = 0.3$ m/s for CH_4 in Fig. 7.9). At high U_{IN} on the other hand, a more effective upstream heat transfer through the solid for high thermal conductivity materials minimizes the light-off distance rendering the reactor less prone to blowout (see, e.g., $U_{IN} = 0.8$ m/s for C_3H_8 in Fig. 7.9).

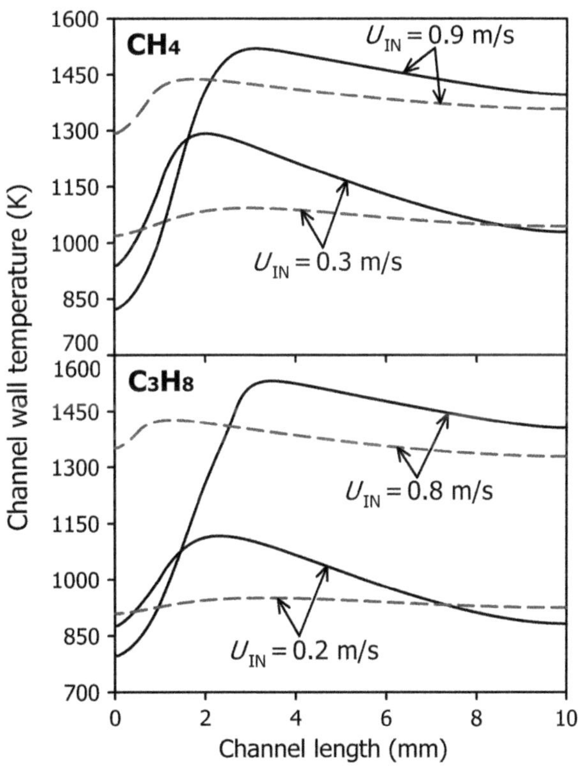

Fig. 7.9 Channel wall temperature profiles of methane- and propane-fueled catalytic microreactors (*top and bottom graphs*, respectively) for two inlet velocities. *Solid lines*: $k_s = 2$ W/mK. *Dashed lines*: $k_s = 16$ W/mK. Other conditions: $p = 5$ bar, $b = 0.15$ mm, $h = 8$ W/m^{2}K

7.4 Conclusions

The combustion stability of propane- and methane-fueled, Pt-coated catalytic microreactors was investigated numerically in a plane channel configuration at 1 and 5 bar, wall thermal conductivities 2 and 16 W/mK, and channel half-heights $b = 0.50$ and 0.15 mm. A 2-D elliptic model was employed, featuring full treatment of all heat transfer mechanisms in the channel reactor, coupled to a detailed surface mechanism for CH_4, a recently proposed global step for C_3H_8 valid for the present moderate pressures and detailed gas-phase chemical reaction schemes for both fuels. Stability diagrams of critical heat transfer coefficient h versus inlet velocity U_{IN} were constructed for both fuels. Propane, similar to methane, substantially benefits from increased operating pressures, due to a positive $p^{+0.75}$ dependence of catalytic reactivity. The impact of gas-phase reactions in extending the blowout limits of propane is more pronounced than in methane, because of the higher gaseous reactivity of the former fuel. Despite the higher catalytic and gas-phase reactivity of C_3H_8 compared to CH_4, the latter exhibits a significantly broader stable combustion regime than the former, allowing higher mass throughputs and external heat losses for the same microreactor configuration. Numerical experiments with an artificial fuel $C_3H_8(*)$ (having the same chemical properties with propane but with $\mathrm{Le} = 1$) revealed that, at higher inlet velocities, the transverse diffusive transport of fuel on the catalytic surface becomes dominant in determining reactor stability, an effect accentuated at elevated pressures ($p = 5$ bar). As the channel half-height decreases from $b = 0.50$ to 0.15 mm, however, the fuel transport properties become less important, due to the associated shorter characteristic mass diffusion time scales. The impact of gas-phase chemistry on reactor stability limits is still significant for the smaller channel gap $b = 0.15$ mm, but less pronounced compared to $b = 0.50$ mm. Finally, as channel confinement increases, the critical heat transfer coefficient h increases for the same mass throughput for sufficiently long residence times, while increasing the solid thermal conductivity from 2 to 16 W/mK widens the stability limits for both propane and methane fuels.

References

1. Isomura K, Murayama M, Teramoto S, Hikichi K, Endo Y, Togo S, Tanaks S (2006) Experimental verification of the feasibility of a 100 W class micro-scale gas turbine at an impeller diameter of 10 mm. J Micromech Microeng 16:S254–S261
2. Peirs J, Waumans T, Vleugels P, Al-Bender F, Stevens T, Verstraete T, Stevens S, D'hulst R, Verstraete D, Fiorini P, Van Der Braembussche R, Driesen J, Puers R, Hendrick P, Baelmans M, Reynaerts D (2007) Micropower generation with microgasturbines: a challenge. J Mech Eng Sci 221:489–500
3. Schneider B, Karagiannidis S, Bruderer M, Dyntar D, Zwyssig C, Guangchun Q, Diener M, Boulouchos K, Abhari RS, Guzzella L, Kolar JW (2005) Ultra-high-energy-density converter for portable power, Power-MEMS 2005, 28–30 Nov 2005, Tokyo, Japan

4. Karagiannidis S, Mantzaras J, Jackson G, Boulouchos K (2007) Hetero-/homogeneous combustion and stability maps in methane-fueled catalytic microreactors. Proc Combust Inst 31:3309–3317
5. Dogwiler U, Benz P, Mantzaras J (1999) Two-dimensional modeling for catalytically stabilized combustion of a lean methane-air mixture with elementary homogeneous and heterogeneous chemical reactions. Combust Flame 116:243–258
6. Schneider A, Mantzaras J, Jansohn P (2006) Experimental and numerical investigation of the catalytic partial oxidation of CH_4/O_2 mixtures diluted with H_2O and CO_2 in a short contact time reactor. Chem Eng Sci 61:4634–4646
7. Siegel R, Howell JR (1981) Thermal radiation heat transfer. Hemisphere, New York, p 271
8. Karagiannidis S, Mantzaras J, Bombach R, Schenker S, Boulouchos K (2009) Experimental and numerical investigation of the hetero-/homogeneous combustion of lean propane/air mixtures over platinum. Proc Combust Inst 32:1947–1955
9. Mantzaras J (2006) In: Jiang SZ (ed) Focus on combustion research. Interplay of transport and hetero-/homogeneous chemistry, Nova Publishers, New York, p 241
10. Appel C, Mantzaras J, Schaeren R, Bombach R, Inauen A, Kaeppeli B, Hemmerling B, Stampanoni A (2002) An experimental and numerical investigation of homogeneous ignition in catalytically stabilized combustion of hydrogen/air mixtures over platinum. Combust Flame 128:340–368
11. Reinke M, Mantzaras J, Schaeren R, Bombach R, Inauen A, Schenker S (2005) Homogeneous ignition of CH_4/air and H_2O- and CO_2-diluted CH_4/O_2 mixtures over platinum; an experimental and numerical investigation at pressures up to 16 bar. Proc Combust Instit 30:2519–2527
12. Qin Z, Lissianski VV, Yang H, Gardiner WC, Davis SG, Wang H (2000) Combustion chemistry of propane: a case study of detailed reaction mechanism optimization. Proc Combust Inst 28:1663–1669
13. Deutschmann O, Maier LI, Riedel U, Stroemman AH, Dibble RW (2000) Hydrogen assisted catalytic combustion of methane on platinum. Catal Today 59:141–150
14. Warnatz J, Dibble RW, Maas U (1996) Combustion, physical and chemical fundamentals, modeling and simulation. Springer-Verlag, New York
15. Reinke M, Mantzaras J, Schaeren R, Bombach R, Inauen A, Schenker S (2004) High-pressure catalytic combustion of methane over platinum: in situ experiments and detailed numerical predictions. Combust Flame 136:217–240
16. Reinke M, Mantzaras J, Bombach R, Schenker S, Inauen A (2005) Gas-phase chemistry in catalytic combustion of methane/air mixtures over platinum at pressures of 1 bar to 16 bar. Combust Flame 141:448–468
17. Kee RJ, Dixon-Lewis G, Warnatz J, Coltrin ME, Miller JA (1996) A fortran computer code package for the evaluation of gas-phase multicomponent transport properties. Report No. SAND86-8246; Sandia National Laboratories
18. Coltrin ME, Kee RJ, Rupley FM (1996) Surface Chemkin: a fortran package for analyzing heterogeneous chemical kinetics at the solid surface-gas phase interface. Report No. SAND90-8003C; Sandia National Laboratories
19. Kee RJ, Rupley FM, Miller JA (1996) Chemkin II: a fortran chemical kinetics package for the analysis of gas-phase chemical kinetics. Report No. SAND89-8009B; Sandia National Laboratories
20. Westbrook CK, Dryer FL (1981) Simplified reaction mechanisms for the oxidation of hydrocarbon fuels in flames. Combust Sci Technol 27:31–43
21. Norton DG, Wetzel ED, Vlachos DG (2004) Fabrication of single-channel catalytic microburners: effect of confinement on the oxidation of hydrogen/air mixtures. Ind Eng Chem Res 43:4833–4840
22. Garetto TF, Rincon E, Apesteguia CR (2004) Deep oxidation of propane on Pt-supported catalysts: drastic turnover rate enhancement using zeolite supports. Appl Catal B Environ 48:167–174

23. Norton DG, Vlachos DG (2004) A CFD study of propane/air microflame stability. Combust Flame 138:97–107
24. Moffat HK, Kee RJ, Grcar JF, Miller JA (1993) Surface PSR: a fortran program for modeling well-stirred reactors with gas and surface reactions. Report No. SAND91-8001; Sandia National Laboratories
25. Kaisare NS, Deshmukh SR, Vlachos DG (2008) Stability and performance of catalytic microreactors: Simulations of propane catalytic combustion on Pt. Chem Eng Sci 63:1098–1116
26. Pfefferle WC, Pfefferle LD (1986) Catalytically stabilized combustion. Prog Energy Combust Sci 12:25–41
27. Satterfield CN, Resnick H, Wentworth RL (1954) Simultaneous heat and mass transfer in a diffusion-controlled chemical reaction-part I: studies in a tubular reactor. Chem Eng Prog 50:460–466

Chapter 8
Numerical Investigation on the Start-Up of Methane-Fueled, Catalytic Microreactors

8.1 Preface

In the present work, a numerical study is undertaken to investigate the start-up of methane-fueled, catalytic microreactors under conditions pertinent to microturbine-based power generation devices. A full elliptic, transient in the solid and quasisteady in the gas numerical code is used to simulate the reacting flow in a catalytic plane channel configuration with a gap of 1 mm and a length of 10 mm, with this setup effectively representing a single channel of a catalytic honeycomb combustor structure. Detailed catalytic and gas-phase reaction mechanisms for the total oxidation of methane on platinum are used. The main objectives are to study the ignition characteristics for microreactors of a nominal geometry by independently varying the pressure, reactor wall material (including wall thermal conductivity and heat capacity), inlet mixture velocity, radiation properties of the solid, and fuel-to-air equivalence ratio. The role of gas-phase chemistry in the start-up process is also investigated. Particular objectives are to assess the impact of the aforementioned parameters on the elapsed time required for reactor ignition and subsequent attainment of steady state.

This chapter is organized as follows. The numerical model is firstly presented, followed by an assessment of characteristic time scales which are relevant for the adopted quasisteady approach. The impact of pressure, solid material properties, equivalence ratio, inlet velocity, surface radiation heat transfer, and gas-phase chemistry on the transient microreactor response is then elaborated. Based on the outcome of the previous computations, implications for the design of microreactor systems are outlined.

S. Karagiannidis, *Catalytic Microreactors for Portable Power Generation*, Springer Theses, DOI: 10.1007/978-3-642-17668-5_8,

8.2 Numerical Model

A full-elliptic, two-dimensional CFD code [1–3] has been used to simulate the laminar flow domain (the maximum inlet Reynolds number being ~ 55 for the cases considered) in a plane channel configuration having length $L = 10$ mm, height $2b = 1$ mm and wall thickness $\delta = 50$ μm (see Fig. 8.1). The initial 1 mm channel length was catalytically inert, while the remaining $L_a = 9$ mm was coated with platinum. Due to symmetry, only half of the channel domain was modeled. The examined fuel-to-air equivalence ratios were $\varphi = 0.4$ and 0.6, while the inlet temperature was set to $T_{IN} = 850$ K, a value practically achievable in recuperated microreactor thermal cycles [4]. The initial temperature for the channel solid wall was uniform and equal to the incoming mixture temperature, such that T_w $(x, t = 0) = 850$ K. Calculations were performed for pressures $p = 1$, 2, 3, 4 and 5 bar, the range 3–5 bar being of particular interest to microturbine-based microreactor systems [4, 5]. The bulk of the present calculations refer to $p = 5$ bar. The nominal inlet velocity was $U_{IN} = 1.5$ m/s at $p = 1$ bar, a value typical for microreactors. When increasing the inlet pressure, the inlet velocity was decreased accordingly, so as to maintain the same mass throughput ($\rho_{IN}U_{IN}$). Two types of solid materials were examined, cordierite (ceramic) and FeCr alloy (metallic), with their respective properties [6] provided in Table 8.1.

The combustion of methane on platinum was modeled using the detailed heterogeneous scheme of Deutschmann et al. [7] (24 reactions, 11 surface and 9 gaseous species), coupled to the C1/H/O elementary gas-phase mechanism of Warnatz et al. [8] (26 species, 108 reactions). In the catalytic mechanism, a surface site density $\Gamma = 2.7 \times 10^{-9}$ mol/cm^2 was used. The catalytic mechanism has been validated against spatially-resolved measurements of major species concentrations across the boundary layer formed in a Pt-coated channel at pressures of 1–16 bar [9].

The gas-phase mechanism has been tested against OH laser induced fluorescence (LIF) homogeneous ignition measurements [10] in the same channel reactor, again at pressures of 1–16 bar. To reproduce homogeneous ignition at $p \leq 6$ bar (a range of particular interest for microreactors), the gaseous mechanism has been modified in the single reaction CHO + M ⇔ CO + H + M; this modification was further supported by recent kinetic measurements (see discussion in [10]). Transport properties were calculated using the CHEMKIN database [11], while

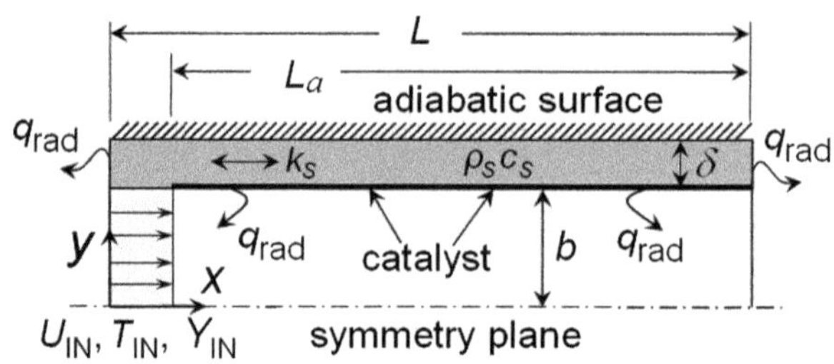

Fig. 8.1 Schematic of the catalytic microreactor configuration

Table 8.1 Properties for cordierite and FeCr alloy channel wall materials

Material	k_s	ρ_s	c_s	$\rho_s c_s$
Cordierite	2.0	2,600	1,464	3,806
FeCr alloy	16.0	7,200	615	4,428

Thermal conductivity k_s (W/mK), density ρ_s (kg/m^3), specific heat capacity c_s (J/kgK) and heat capacity $\rho_s c_s$ (kJ/m^{3}K)

surface and gas-phase reaction rates were evaluated using Surface-CHEMKIN [12] and CHEMKIN, respectively [13].

All channel elements had an emissivity $\varepsilon_j = \varepsilon = 0.6$, $j = 1, \ldots, N$, while examples of the calculated factors $F_{k\text{–}j}$ are presented in Fig. 8.2 for two channel wall elements and the inlet channel enclosure. The inlet and outlet planes of the enclosure had emissivities equal to those of the channel wall surfaces, $\varepsilon_{IN} = \varepsilon_{OUT} = \varepsilon = 0.6$, while the inlet and outlet exchange temperatures were set equal to the inlet mixture and outlet mixing cup temperatures, respectively. This arrangement mimics the tight space in microreactor systems, wherein the entry and outlet sections cannot usually be of large enough size to allow for a black body enclosure treatment. The outer horizontal wall of the microreactor channel was treated as adiabatic (see Fig. 8.1); nevertheless, the reactor itself was non-adiabatic due to radiation heat losses, primarily from the channel wall inner surface as well as from the vertical front solid wall face towards the colder inlet enclosure.

An orthogonal staggered grid with 100 × 24 points in x- and y-direction, respectively, over the channel half-height produced a grid-independent solution for the flow domain. Finer spacing towards the channel wall and entry section was used. A 100 grid node resolution in the x-direction was also used to discretize the solid wall. At the inlet ($x = 0$), uniform profiles of species, temperature and axial velocity were applied, while zero-Neumann conditions were set at the outlet ($x = L$) and plane of symmetry ($y = 0$). No-slip was applied for both velocity components at the gas-wall interface ($y = b$).

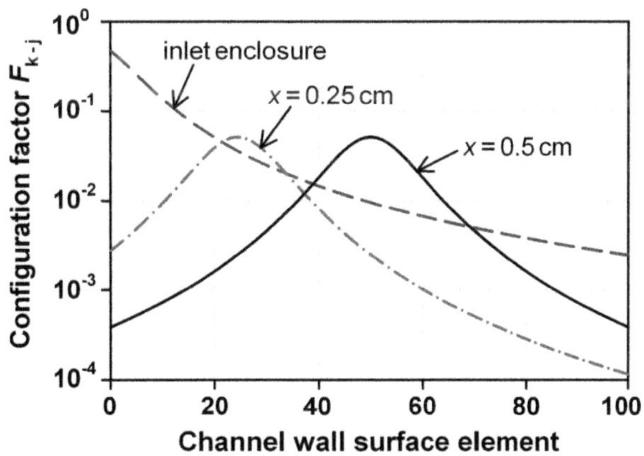

Fig. 8.2 Configuration factors $F_{k\text{-}IN}$ between each channel wall surface element k and the inlet (*dashed line*), $F_{k\text{–}j}$ between the k-th wall element at $x = 0.25$ cm and the remaining wall elements j (*dash-dotted line*), and finally $F_{k\text{–}j}$ between the k-th wall element at $x = 0.50$ cm and the remaining j wall elements (*solid line*)

The quasisteady assumption requires that the convective and diffusive time scales of the gas are appreciably shorter than the characteristic time scales for diffusion of heat in the solid, thus allowing the gaseous flow to equilibrate to the channel solid wall temperature at any given time during the ignition event [1, 14]. The 2-D steady calculations of the flow field were thus coupled to a 1-D transient energy balance equation for the solid. The choice of a 1-D model for the solid energy equation will be clarified in Sect. 8.3.1. The quasisteady approximation also entails the assumption of catalytic chemical reaction times being shorter than the heat conduction times in the solid, so as to ensure chemical equilibration at the local wall temperature during an integration time step of the solid. The time step $\Delta t = 50$ ms used in this work was thus longer than the chemical time scales present during catalytic microreactor ignition (discussion on the chosen time step is provided in Sect. 8.3.1).

All computations were performed on a cluster with 32 Xeon CPUs at 3.0-GHz. Computations were very demanding and a time integration of 30 s required up to $\sim$ 15 days on a single CPU when full hetero-/homogeneous chemistry was included. These CPU time requirements clearly show why fully transient (for both the gas- and solid-phase processes) multidimensional models with realistic hetero-/ homogeneous chemistry have not so far been presented in the literature.

8.3 Results and Discussion

Transient simulations were performed in order to assess the impact on the microreactor start-up of various operating parameters, such as the inlet pressure p, inlet flow velocity U_{IN}, methane-to-air equivalence ratio φ, solid wall properties k_s and $\rho_s c_s$, emissivity of channel surfaces ε, and finally gas-phase reactions. The computed conditions are summarized in Table 8.2.

Of main interest was to assess the effect of the aforementioned parameters on two characteristic times that describe the microreactor start-up. The ignition time (t_{ig}) was defined as the elapsed time required to reach 50% of methane conversion at the channel outlet, and the steady-state time (t_{st}) as the elapsed time whereby the outlet gas temperature varied by less than 10^{-3} K. The former definition was meaningful, since complete methane conversion (> 99.99%) could be achieved under catalytic mass-transport-limited steady operation for all conditions examined. Moreover, by running a steady-state version of the code [2], it was confirmed that the adopted definition of steady state in the transient simulations reproduced the true steady-state outlet temperature within 1 K.

In the following sections, the impact of microreactor operating conditions and wall material on the large variation of characteristic start-up times (see Table 8.2) will be addressed, along with a characteristic time scale analysis identifying relevant time scales.

Table 8.2 Numerical conditions and results

Case	Material	p	φ	U_{IN}	ε	ε_{IN}, ε_{OUT}	t_{ig}	t_{st}
1	Cordierite	1	0.4	1.50	0.6	0.6	24.3	41.1
2	Cordierite	2	0.4	0.75	0.6	0.6	15.6	31.6
3	Cordierite	3	0.4	0.50	0.6	0.6	13.5	28.9
4	Cordierite	4	0.4	0.38	0.6	0.6	12.6	27.7
5	Cordierite	5	0.4	0.30	0.6	0.6	11.7	25.8
6	FeCr alloy	1	0.4	1.50	0.6	0.6	26.6	41.7
7	FeCr alloy	2	0.4	0.75	0.6	0.6	19.1	34.4
8	FeCr alloy	3	0.4	0.50	0.6	0.6	16.9	32.0
9	FeCr alloy	4	0.4	0.38	0.6	0.6	15.9	30.9
10	FeCr alloy	5	0.4	0.30	0.6	0.6	15.0	28.8
11	Cordierite	5	0.6	0.30	0.6	0.6	5.4	16.9
12	FeCr alloy	5	0.6	0.30	0.6	0.6	6.8	19.1
13	Cordierite	5	0.6	0.50	0.6	0.6	7.1	15.3
14	FeCr alloy	5	0.6	0.50	0.6	0.6	9.0	17.5
15	Cordierite	5	0.4	0.30	0.0	0.0	11.0	29.6
16	Cordierite	5	0.4	0.30	0.6	0.0	11.2	26.8
17	Cordierite*	5	0.4	0.30	0.0	0.0	11.2	26.5
18	FeCr alloy	5	0.4	0.30	0.0	0.0	13.9	29.4
19	FeCr alloy	5	0.4	0.30	0.6	0.0	13.9	29.5
20	Cordierite[a]	1	0.4	1.50	0.6	0.6	24.0	42.0
21	Cordierite[a]	5	0.4	0.30	0.6	0.6	11.5	26.4
22	FeCr alloy[a]	1	0.4	1.50	0.6	0.6	26.1	42.8
23	FeCr alloy[a]	5	0.4	0.30	0.6	0.6	14.8	31.2
24	Cordierite[a]	5	0.6	0.30	0.6	0.6	5.4	18.3
25	FeCr alloy[a]	5	0.6	0.30	0.6	0.6	6.8	22.4

Reactor material, inlet pressure p (bar), equivalence ratio φ, inlet velocity U_{IN} (m/s), surface emissivity ε, inlet/outlet enclosure emissivity $\varepsilon_{IN}/\varepsilon_{OUT}$, ignition time t_{ig} (s) and steady-state time t_{st} (s). Cases 1–19 pertain to simulations with surface reactions only. Case 17: cordierite* denotes a material with thermal conductivity $k_s = 5.5$ W/mK

[a] Cases 20–25: simulations with catalytic and gas-phase chemistry

8.3.1 Characteristic Time Scale Analysis

Under the quasisteady formulation of Eqs. 3.1–3.8 and 3.11, the characteristic time scales for heat conduction in the solid substrate must be sufficiently longer than the characteristic convective, diffusive, and chemical time scales of the reacting flow inside the channel. This ensures that at every time step the gaseous flow and chemistry equilibrate to the imposed, at every time step, solid wall temperature. Accordingly, the time step Δt must be longer than the characteristic convective, diffusive and chemical time scales of the reacting flow, but short enough to accurately resolve the transient heat response of the solid.

For the axial convection inside the microreactor, the characteristic time $t_{g,x} \approx L/U_{IN}$ (not accounting for flow acceleration of the gas due to combustion) was estimated between 6 and 33 ms. Diffusive transport time scales of the gas in the y-direction were estimated as $t_{g,y} \approx b^2/a_g$, with $a_g = k/(\rho c_p)$ the thermal diffusivity of the gas calculated for pressures $p = 1$–5 bar, and temperatures of 850–1,850 K (minimum and maximum gas temperatures in the microreactor, as will be shown in the forthcoming sections); the calculations yielded $t_{g,y} \approx 0.5$–2.0 ms for $p = 1$ bar and ~ 2.6–10.2 ms for $p = 5$ bar. On the other hand, the corresponding times for solid substrate heat conduction in the x- and y-direction were $t_{s,x} \sim L^2/a_s$ and $t_{s,y} \sim \delta^2/a_s$, with $a_s = k_s/(\rho_s c_s)$ the solid thermal diffusivity. For cordierite, these characteristic times assumed values of $t_{s,x} \approx 190.3$ s and $t_{s,y} \approx 4.7$ ms, while for the FeCr alloy $t_{s,x} \approx 27.6$ s and $t_{s,y} \approx 0.7$ ms. Since $t_{s,x} \gg t_{g,x}, t_{g,y}$ for both materials, the quasisteady assumption could be safely invoked. However, given that for FeCr alloy the characteristic time for transverse solid heat conduction $t_{s,y}$ was in most cases much shorter than the calculated times $t_{g,y}$, the time evolution of the transverse temperatures inside the 50 μm thick solid could not be resolved; therefore, a 1-D axial energy balance has been employed for the solid wall in Eq. 3.11. For consistency, and since for both wall materials $t_{s,x} \gg t_{s,y}$, the same approach was adapted for cordierite, even though for this material $t_{s,y} > t_{g,y}$, at least for $p = 1$ bar.

Characteristic chemical time scales for lean CH_4/air catalytic combustion on Pt also had to be considered, since they could be quite long before and during light-off, especially if the initial surface and gas temperatures were relatively low. It has been pointed out [14] that the quasisteady assumption, although widely used in many catalytic applications, may be invalidated at specific times and/or spatial locations of the reactor, whereby the characteristic catalytic chemical times become longer than the solid heat conduction times. For the present operating conditions, however, the high initial solid temperature (850 K) largely removes such concerns, as shown next. A constant-pressure batch reactor model, with both catalytic and gas-phase reactions, was used to assess characteristic chemical time scales for all reacting gaseous species. The numerical code has been constructed by augmenting the gas-phase batch reactor package SENKIN of CHEMKIN [15] (model details have been provided in [1]). The $p = 1$ bar and $\varphi = 0.4$ case was deemed to pose the most stringent limitations in the ensuing calculations (the catalytic reactivity increases with both increasing methane concentration [16] and pressure [9]). In addition, calculations with an initial batch reactor temperature $T_{IN} = 850$ K provided strict upper estimates for the chemical times, as they mimicked conditions at $t = 0$ in the channel reactor; at later times, the heat up of the channel wall and gas would imply even higher inlet temperatures for the batch reactor simulations. To obtain a more complete picture of the temperature effects, the initial batch reactor temperature was further varied between 750 and 850 K. A surface-to-volume ratio of $S/V = 20$ cm^{-1} was used for the batch reactor, which equals the S/V of the channel geometry in Fig. 8.1. Two cases were studied, one considering only catalytic reactions and the other with both catalytic and gas-phase chemistry. With reactor ignition defined as the elapsed time until 50% of the incoming fuel is consumed, characteristic chemical time scales for each gas-phase species were defined as:

$$\tau_{\mathrm{CH},k} = \frac{1}{t_{ig}} \int_0^{t_{ig}} \frac{C_k}{\dot{s}_k(S/V) + \dot{\omega}_k} dt. \tag{8.1}$$

Chemical time scales for all reacting gas-phase species are presented in Fig. 8.3 for the case of catalytic reactions only, at various batch reactor inlet temperatures. With the characteristic times of gas-phase species ranging from 105 ms (CO) to 476 ms (O_2) for an initial reactor temperature of 750 K, a time step $\Delta t \sim 500$ ms for the time-dependent energy equation of the solid would be required to ensure that all gaseous species equilibrate to the given surface temperature. Due to the relatively short characteristic axial heat conduction time of the FeCr alloy ($t_{\mathrm{s,x}} \sim 27$ s), such a high Δt would entail poor description of the time evolution of the wall temperature during the heat-up phase. A favorable choice for microreactor inlet temperature and time step is $T_{\mathrm{IN}} = 850$ K and $\Delta t = 50$ ms, respectively; this choice is consistent with the quasisteady gas phase assumption and, moreover, provides good resolution of the solid temperature time evolution.

In a further analysis, the same characteristic time scale analysis was performed as before, but with the inclusion of gas-phase reactions. The calculated characteristic chemical time scales are presented in Fig. 8.4. Important intermediate species of lean CH_4/air combustion such as formaldehyde (CH_2O) and acetylene (C_2H_2) now impose further restrictions on the required time step Δt, as they have characteristic times of the same magnitude as total oxidation products CO_2 and H_2O. However, a significant drop of the chemical times to below 50 ms is evident in Fig. 8.4 for all species at $T_{\mathrm{IN}} = 850$ K. This further justifies the use of the aforementioned combination of Δt and T_{IN} for all subsequent numerical simulations. It is finally emphasized that the analysis in Figs. 8.3 and 8.4 is quite strict when applied to the channel in Fig. 8.1, since as the solid starts heating above the initial temperature $T_w(x, t = 0) = 850$ K, the chemical time scales shorten substantially: already at 900 K, the chemical times are a factor of $\sim$ 2.2–3.5 shorter than the ones at 850 K shown in Figs. 8.3 and 8.4.

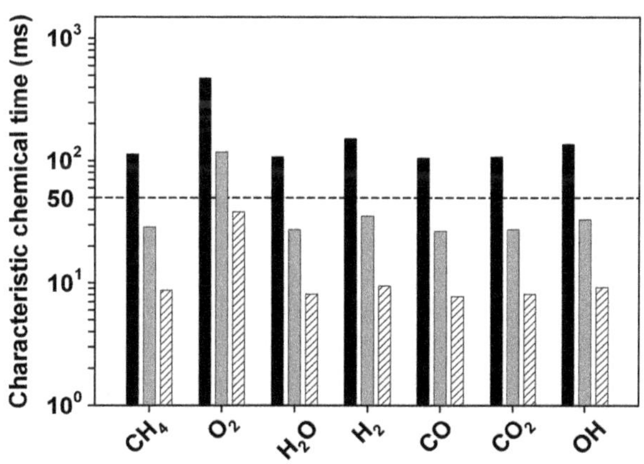

Fig. 8.3 Computed characteristic catalytic chemical time scales of gas-phase reacting species in a batch reactor, when considering only catalytic reactions. Reactor conditions: $p = 1$ bar, $\varphi = 0.4$, $S/V = 20$ cm^{-1}. *Black bars*: $T_{\mathrm{IN}} = 750$ K, *gray bars*: $T_{\mathrm{IN}} = 800$ K, *dashed bars*: $T_{\mathrm{IN}} = 850$ K. *Dashed line*: time step $\Delta t = 50$ ms used in subsequent channel-flow simulations

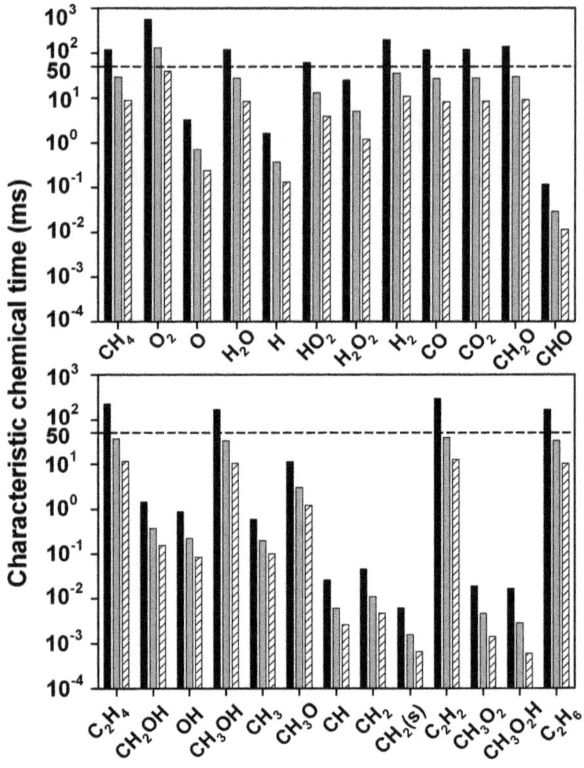

Fig. 8.4 Computed characteristic chemical time scales of gas-phase reacting species in a batch reactor, when considering both catalytic and gas-phase reactions. Reactor conditions: $p = 1$ bar, $\varphi = 0.4$, $S/V = 20\ \text{cm}^{-1}$. *Black bars*: $T_{IN} = 750$ K, *gray bars*: $T_{IN} = 800$ K, *dashed bars*: $T_{IN} = 850$ K. *Dashed line*: time step $\Delta t = 50$ ms used in subsequent channel-flow simulations

8.3.2 Effect of Microreactor Inlet Pressure

Elevated operating pressures have been shown to substantially extend the stable combustion regime of methane-fueled catalytic microreactors under steady-state operation [17], the reason being that the catalytic reactivity of CH_4 on Pt follows a positive $p^{+0.47}$ dependence on pressure [9]. A reduction in the ignition (t_{ig}) and steady-state (t_{st}) times is thus expected with rising pressure. This effect is illustrated in Fig. 8.5, where computed t_{ig} and t_{st} times are plotted for Cases 1–10.

The mass throughput, $\rho_{IN}U_{IN}$, was held constant as the inlet pressure increased by reducing accordingly the inlet velocity U_{IN}. This facilitated comparisons between different cases, since analytical studies have shown that the catalytic fuel conversion is a sole function of the inlet Reynolds number—at least when all other parameters (geometrical, transport and kinetic) are kept the same [18]. For both microreactor wall materials investigated, a fivefold increase in inlet pressure decreased by almost a factor of two the ignition times. Specifically, for cordierite the reduction in t_{ig} was ~ 51% ($t_{ig} = 24.3$ s at $p = 1$ bar, against $t_{ig} = 11.7$ s at $p = 5$ bar), while for FeCr alloy ~ 43% ($t_{ig} = 26.6$ s at $p = 1$ bar, against $t_{ig} = 15.0$ s at $p = 5$ bar in Fig. 8.5).

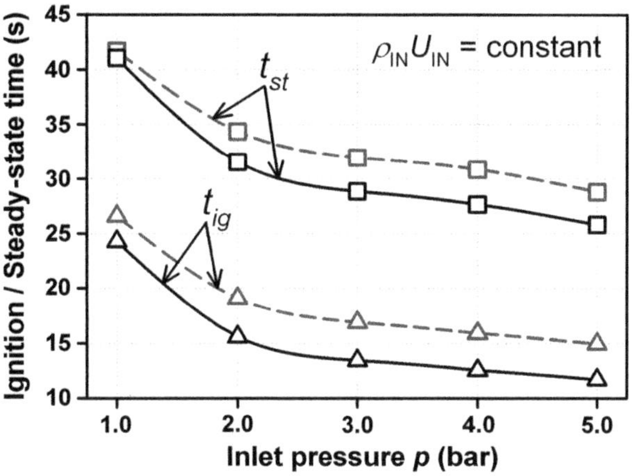

Fig. 8.5 Ignition (t_{ig}) and steady-state (t_{st}) times versus inlet pressure for Cases 1–10 in Table 8.2. *Triangles*: ignition times; *squares*: steady-state times. *Solid lines*: cordierite reactor; *dashed lines*: FeCr alloy reactor. The mass inflow ($\rho_{IN}U_{IN}$) is constant for all cases

The increased catalytic reactivity of methane on Pt at elevated pressures allows for significant fuel consumption at lower wall temperatures than those required at atmospheric pressure, thus facilitating an earlier microreactor ignition. This is evidenced in Figs. 8.6 and 8.7, where streamwise methane catalytic conversion rates and channel wall temperature profiles are plotted for Cases 1 and 5, at $p = 1$ and 5 bar, respectively.

In comparison to Case 1, Case 5 exhibits more vigorous combustion at all times, from ignition till steady state (see Fig. 8.6). Both cases display a rear-end ignition, with the reaction zone gradually propagating upstream. With the propagation speed of the reaction front being controlled by the solid thermal diffusivity

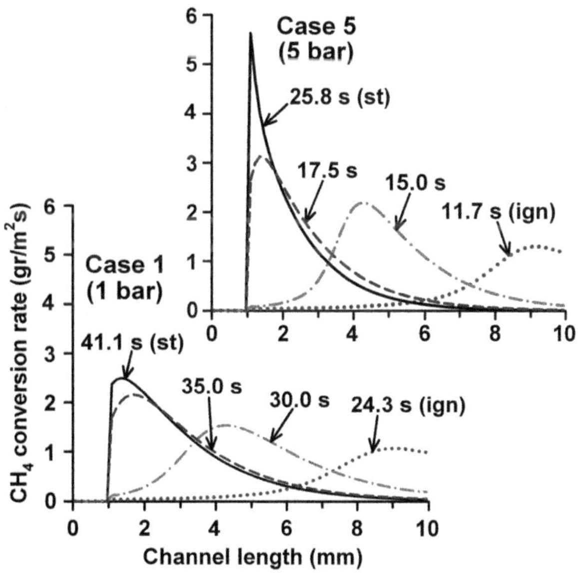

Fig. 8.6 Methane catalytic conversion rates along the microreactor during the start-up phase for Cases 1 (*bottom graph*) and 5 (*top graph*), at 4 time instances: ignition (ign), steady state (st) and two intermediate times. Cordierite wall material

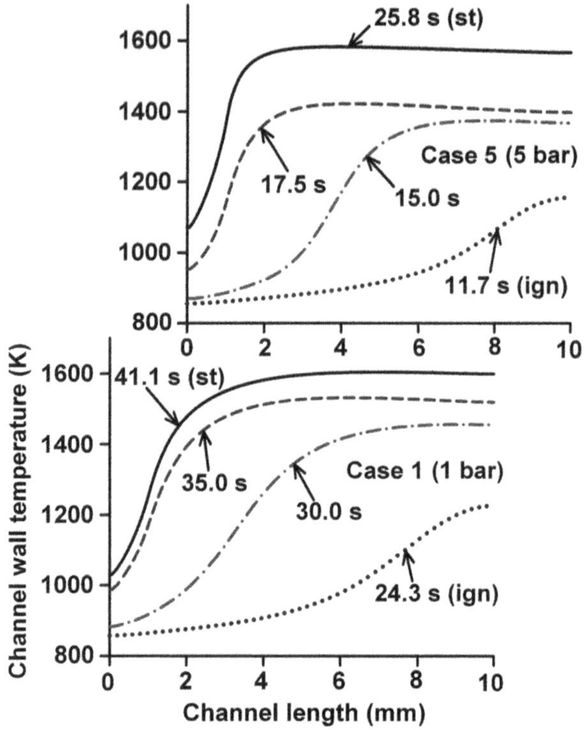

Fig. 8.7 Channel wall temperature profiles during the start-up phase for Cases 1 (*bottom graph*) and 5 (*top graph*), at 4 time instances: ignition (ign), steady state (st) and two intermediate times. Cordierite wall material

(a_s) and the local heat release ($\propto \dot{s}_k$) [19], the reaction zone reaches the beginning of the coated section ($x = 1$ mm) after ignition much faster in Case 5 compared to Case 1 (in 5.8 and 10.7 s following $t_{ig} = 11.7$ and 24.3 s, respectively, see Fig. 8.6) resulting also in a faster wall temperature rise (Fig. 8.7). However, this difference does not result in a significantly faster approach to steady state for the higher pressure case: it takes 14.1 s for Case 5 to reach steady state after ignition, as compared to 16.8 s for Case 1. This indicates that once the microreactor is ignited throughout its length, the chemical power input is the controlling factor in how fast the steady state is approached.

It would appear that microreactors with higher wall heat capacity (such as FeCr alloy) may be more significantly affected by operating pressure during start-up, since they would otherwise require a longer time to heat up to the critical temperatures necessary for catalyst light-off. A study of the results in Table 8.2, however, reveals an opposite trend. While for a cordierite microreactor an increase from 1 to 5 bar induces a reduction of t_{ig} and t_{st} by ~ 51% and ~ 37%, respectively (Cases 1 and 5), for a FeCr alloy the reduction in the corresponding characteristic times is ~ 43% and ~ 30% (Cases 6 and 10). This behavior is clarified with the aid of Fig. 8.8, providing transverse CH_4 mol fraction profiles across the channel half-height at 5 selected streamwise positions for Cases 1, 6, 5 and 10 at ignition, steady state and two intermediate times. The boundary layer

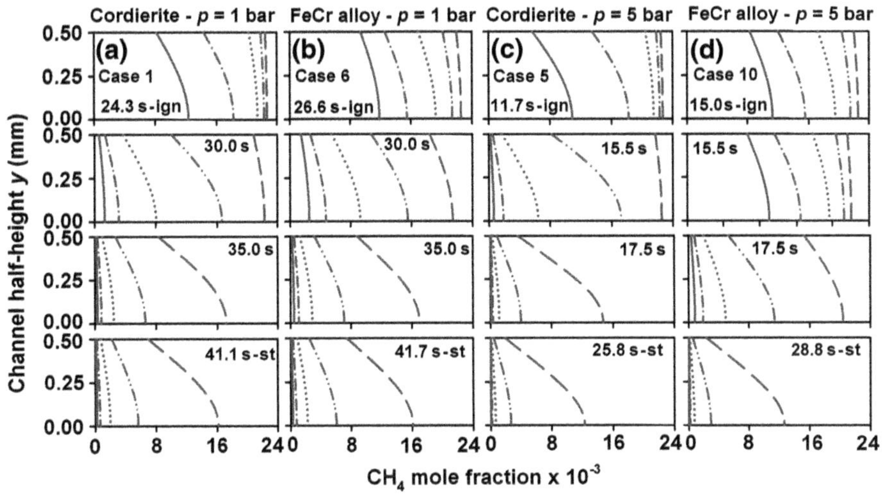

Fig. 8.8 Transverse ($y = 0.5$ mm is the channel wall and $y = 0$ the symmetry plane) methane mole fractions for Cases 1, 6, 5 and 10 ((a)–(d), respectively) during ignition (ign), two intermediate time instances and steady state (st), at five streamwise positions: $x = 1.98$ mm (*dashed lines*), 4.03 mm (*dashed-double-dotted lines*), 5.96 mm (*dotted lines*), 8.01 mm (*dashed-dotted lines*) and 9.93 mm (*solid lines*)

profiles of methane indicate that during the ignition and the subsequent heat-up phase at $p = 1$ bar, the fuel conversion is essentially kinetically controlled for both cordierite and FeCr alloy microreactors (manifested by the large CH_4 concentration near the wall over most of the reactor length). At $p = 5$ bar, however, the cordierite microreactor transitions to a fuel conversion that is close to the transport limit much earlier than the FeCr alloy (compare the CH_4 profiles at 17.5 s in Fig. 8.8c, d). As a result, while at 1 bar both cordierite and FeCr alloy microreactors have comparable t_{st} times, at 5 bar the difference increases favoring the microreactor with a cordierite wall. It will be shown in the coming section that this outcome is due to the higher thermal conductivity of the FeCr alloy that in turn leads to a spatially more uniform wall temperature profile and thus to locally lower surface reaction rates.

In summary, by increasing the operational pressure from 1 to 5 bar the ignition characteristics of methane-fueled catalytic microreactors are significantly improved. This result, in conjunction with the broadening of the steady-state combustion stability limits with rising pressure [17], clearly points to the advantages of microreactor operation at moderate pressures. The same result is also expected to hold for other lower hydrocarbons, given their positive catalytic reactivity pressure dependence (i.e. propane [20]). Moreover, low thermal conductivity channel walls provide shorter ignition and steady-state times compared to high thermal conductivity walls, and this effect is more pronounced at higher pressures.

It is finally noted that an increase in catalytic reactivity (and thus a reduction of the ignition and steady-state times) can also be achieved in practice by increasing the catalytically active-to-geometrical surface area ratio, e.g. by applying a catalyst-loaded washcoat on the microreactor channel walls. A limited number of calculations were performed with the active-to-geometrical surface area ratio set to a value of 5 for selected cases in Table 8.2; this value serves as a multiplicative factor for the catalytic rates $\dot{s}_k$ in Eq. 3.15. Characteristically, when recomputing Case 10 in Table 8.2 with the aforementioned higher active-to-geometrical surface area ratio, the microreactor ignition time was reduced to 3.8 s (as opposed to $t_{ig} = 15.0$ s) and the time required to reach steady state dropped to 14.8 s (as opposed to $t_{st} = 28.8$ s). A detailed quantification of the impact of the active-to-geometrical surface area ratios on microreactor start-up is, however, outside the scope of this work.

8.3.3 Impact of Channel Wall Material Properties

In order to study the effect of wall thermal conductivity and heat capacity on the start-up process of catalytic microreactors, two types of representative reactor wall materials have been investigated (see Table 8.1). At steady-state operation, the dominant heat transfer mechanism is the conduction in the channel walls, directly impacting combustor stability by preheating the incoming fresh combustible mixture. During the heat-up process, the heat capacity of the solid plays an important role, since materials with a higher $\rho_s c_s$ would need larger energy input to raise their temperature. Figure 8.9 illustrates the key differences between cordierite and FeCr alloy channel walls.

Wall temperature profiles are presented for Cases 5 and 10, corresponding to microreactors with cordierite and FeCr alloy materials. In Case 5, rear-end ignition is observed, with the catalytic ignition (light-off) distance decreasing with increasing time until steady state is reached. In contrast, Case 10 displays more evenly distributed spatial wall temperature profiles, a behavior in agreement with earlier observations in microburner propane ignition [21]. An appreciable difference in characteristic start-up times is observed between the two materials, with cordierite exhibiting both shorter t_{ig} and t_{st} compared to FeCr alloy, by 3.3 and 3.0 s respectively.

The more favorable start-up times for cordierite are mainly attributed to its lower thermal conductivity. Before ignition, axial heat conduction in the solid is less pronounced for the ceramic material, due to its lower k_s. Heat generated on the surface cannot diffuse away from the reaction front located near the channel exit at a fast enough rate; this leads to the formation of a spatially confined reaction zone (see in Fig. 8.9a the more pronounced hot spot at the reactor rear-end), which in turn promotes faster fuel consumption and leads to faster light-off. This faster light-off is also attributed to the fact that less heat is accumulated in the cordierite compared to the FeCr alloy. In Fig. 8.10, thermal power generated by surface

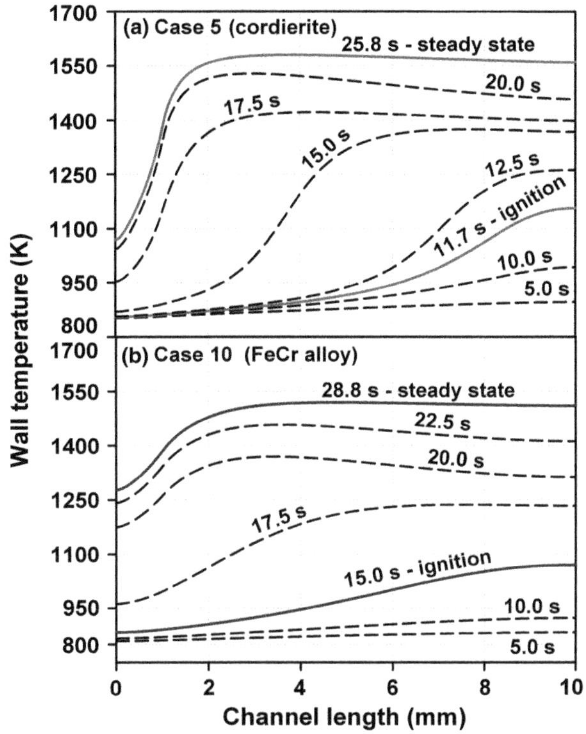

Fig. 8.9 Streamwise wall temperature profiles at ignition and steady state (*solid lines*) and at various other time instances (*dashed lines*) for (**a**) Case 5 and (**b**) Case 10. Other conditions: $p = 5$ bar, $U_{IN} = 0.3$ m/s, $\varphi = 0.4$, $\varepsilon = 0.6$

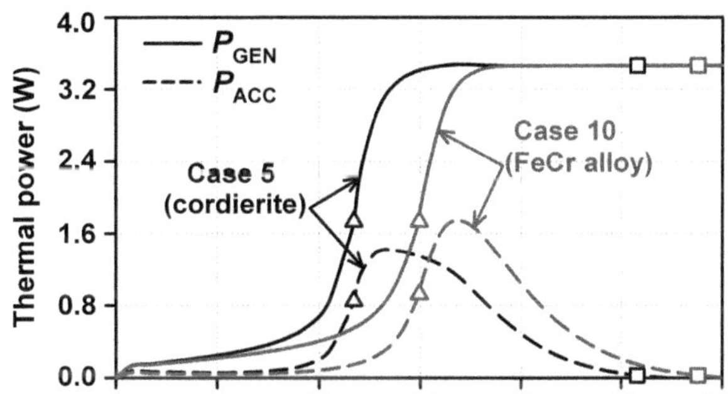

Fig. 8.10 Thermal power generated by surface reactions (P_{GEN}, *solid lines*) and power accumulated in the solid substrate (P_{ACC}, *dashed lines*) summed over all channel wall elements, from $t = 0$ up to t_{st} for Cases 5 and 10. Triangles: t_{ig}. *Squares*: t_{st}. A channel width of 1 cm is considered

reactions (P_{GEN}) and power accumulated by the solid (P_{ACC}) during start-up, summed over all channel wall elements, are presented for Cases 5 and 10, assuming a reactor width of 1 cm (z-direction).

While at steady state in both cases the same thermal power output is achieved (> 99.99% fuel conversion), in Case 5 (cordierite) a smaller amount of energy is stored in the reactor wall as sensible heat. By integrating the P_{ACC} curves in Fig. 8.10 until t_{st}, it is deduced that for cordierite 11.8 J are required to raise the reactor temperature to its steady-state value, while for FeCr alloy the corresponding energy is 12.5 J. However, as will be clarified in Sect. 8.3.5, radiation losses are more pronounced in the metallic wall reactor compared to the ceramic one due to its higher upstream wall temperatures. Consequently, even though the FeCr alloy microreactor has $\sim$ 16.3% higher solid heat capacity than the cordierite one (see Table 8.1), by equilibrating to lower steady-state temperatures the FeCr alloy microreactor requires only $\sim$ 6% more energy to reach steady state.

It should be emphasized that the aforementioned observations regarding material behavior are in stark contrast to steady-state catalytic microreactor results [17]. At steady state, FeCr alloy microreactors provided more robust combustion (extended stability limits) against imposed external heat losses when compared to ceramic ones. On the other hand, the present transient studies clearly show the advantage of ceramic materials in terms of start-up. However, in either steady or transient operation, cordierite microreactors lead to higher wall temperatures (see Fig. 8.9). Both facts must be considered when selecting appropriate reactor materials for specific applications.

8.3.4 Effect of Equivalence Ratio and Inlet Velocity

Calculations were performed for both cordierite and FeCr alloy wall materials, at increased inlet velocities of $U_{IN} = 2.50$ and 0.50 m/s and pressures $p = 1$ and 5 bar, respectively, for $\varphi = 0.4$. For time integration up to $t = 150$ s, ignition could not be achieved for both investigated materials. This necessitated an increase in the equivalence ratio to $\varphi = 0.6$ to enhance the catalytic reactivity and thus ignite the microreactor.

Cases 11–14 in Table 8.2 pertain to simulations at $\varphi = 0.6$ and $p = 5$ bar, for inlet velocities $U_{IN} = 0.30$, and 0.50 m/s. By comparing ignition and steady-state times of Cases 11 with 5 and 12 with 10, a clear benefit is evident for the higher φ, with t_{ig} reduced to less than 50% of its corresponding value at $\varphi = 0.4$, for both materials.

This behavior can be explained by examining the surface coverage, which is provided in Fig. 8.11 for Cases 5 and 11, at ignition and 1 s after ignition. Once ignited, the higher φ case exhibits stronger combustion, which is responsible for the marked difference in O(s) and Pt(s) coverage after ignition: the higher free surface site coverage, Pt(s), for the higher equivalence ratio case (particularly for $x > 4$ mm) manifests enhanced catalytic methane conversion rates (methane adsorption, which is the limiting step in the surface reaction mechanism scales as $Pt(s)^{2.3}$ [7, 9]).

As a result of the increased catalytic reactivity of methane on Pt with rising equivalence ratio (see also [16]), the differences in characteristic times between cordierite and FeCr alloy, and in particular the ignition times, diminish. Direct comparisons for both materials and equivalence ratios are provided in Fig. 8.12, where the exhaust gas temperature of the microreactor is plotted versus the elapsed

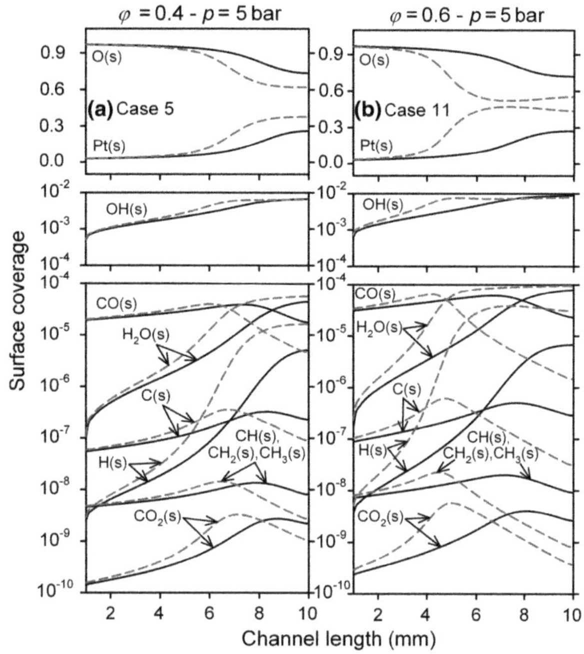

Fig. 8.11 Surface coverage at ignition (*solid lines*) and 1 s after t_{ig} (*dashed lines*) for (**a**) Case 5 and (**b**) Case 11. The coverage is provided for the catalytically coated section $x \geq 1$ mm. Cordierite wall material

time, for Cases 5, 10, 11 and 12. For both equivalence ratios, the outlet gas temperature for a cordierite microreactor increases rapidly upon ignition and reaches steady state in a shorter time compared to the FeCr alloy. Moreover, the reduction in ignition times at $\varphi = 0.4$ between the two solid materials, is more pronounced than the corresponding reduction at $\varphi = 0.6$. Thus, the larger catalytic reactivity (and also the larger exothermicity) of the higher φ cases reduces the impact of material properties.

The exhaust gas temperature is substantially lower than the adiabatic flame temperature (see Fig. 8.12), even though at steady state almost complete fuel conversion (> 99.999%) is attained in all cases. The radiation exchange with the inlet section is responsible for significant heat losses, a phenomenon particularly enhanced in catalytic microreactors due to their short length and the resulting large viewing factors between the hot reacting wall elements and the colder inlet enclosure. Based on calculations of the enthalpic content of the exhaust gases, the lowest and highest radiation losses are obtained for Case 5 (cordierite, $\varphi = 0.4$, radiation losses comprise 19.9% of the inlet chemical energy), and Case 12 (FeCr alloy, $\varphi = 0.6$, with a corresponding fraction of 28.8%). Since radiation exchange is strongly dependent on the channel wall temperature (particularly near the front-end section), which is in turn a function of the equivalence ratio, radiation losses are more pronounced at $\varphi = 0.6$. For cordierite, this effect is attenuated due to the lower front-end wall temperatures owing to its lower k_s (see Fig. 8.9). The higher exhaust gas temperatures for cordierite at steady state (see Fig. 8.12) are the result of lower radiation losses, for both equivalence ratios considered.

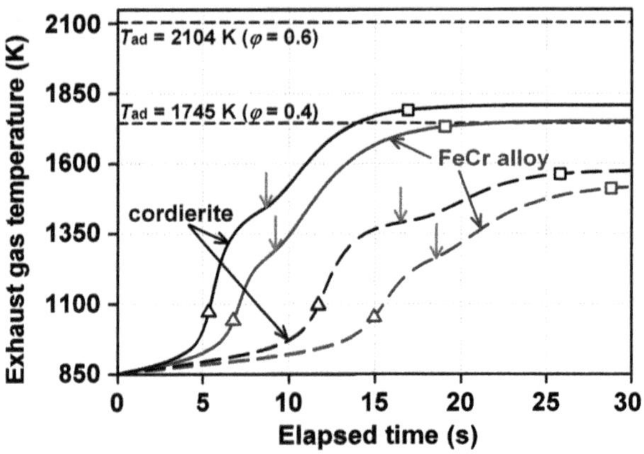

Fig. 8.12 Exhaust gas temperature versus elapsed time for Cases 5, 10, 11 and 12. *Solid lines*: $\varphi = 0.6$. *Dashed lines*: $\varphi = 0.4$. Ignition (t_{ig}) and steady state (t_{st}) times are denoted with *triangles* and *squares*, respectively. *Horizontal dashed lines* correspond to the adiabatic equilibrium temperature at $\varphi = 0.6$ and $\varphi = 0.4$. *Vertical arrows*: time instance at which the reaction front has reached the beginning of the coated section ($x = 1$ mm)

Characteristically, at steady state and $\varphi = 0.4$ radiation losses amount to 25.7% of the enthalpic inlet content for FeCr alloy (Case 10) as opposed to 19.9% for cordierite (Case 5).

The observed changes in the slope of the rising exhaust gas temperature (marked with arrows in Fig. 8.12) occur approximately at the times when the propagating reaction zone reaches the beginning of the coated section ($x = 1$ mm), which locally increases the wall temperature and consequently the radiation losses. The impact of radiation, an important effect neglected in microreactor studies, will be further elaborated in the coming section.

The effect of increased inlet velocity U_{IN} can be evaluated by comparing Cases 11 with 13 and 12 with 14 (see Table 8.2). In Fig. 8.13, channel wall temperature profiles are provided for the aforementioned cases at ignition and steady state. The inlet velocity has a direct impact on both times, since it is directly proportional to the incoming chemical power and the residence time of the fuel/air mixture in the microreactor.

Two opposing trends are seen in Fig. 8.13 with regard to the dependence of t_{ig} and t_{st} on U_{IN}. At $U_{IN} = 0.30$ m/s, the microreactor benefits from an increased flow residence time. This facilitates a faster build-up of heat on the reactor wall and a subsequent faster temperature rise, which leads to a reduced ignition time compared to $U_{IN} = 0.50$ m/s, a fact true for both materials studied. However, once ignited, the higher velocity cases require a shorter time to reach steady state, since they have larger heat generation per unit time that compensates for heat accumulation in the microreactor walls and for radiation losses.

Even though increasing U_{IN} lowers t_{st} for both materials examined (keeping all other parameters the same), a penalty is paid by a substantial increase in the cumulative methane emissions of the catalytic microreactor, since the pre-ignition

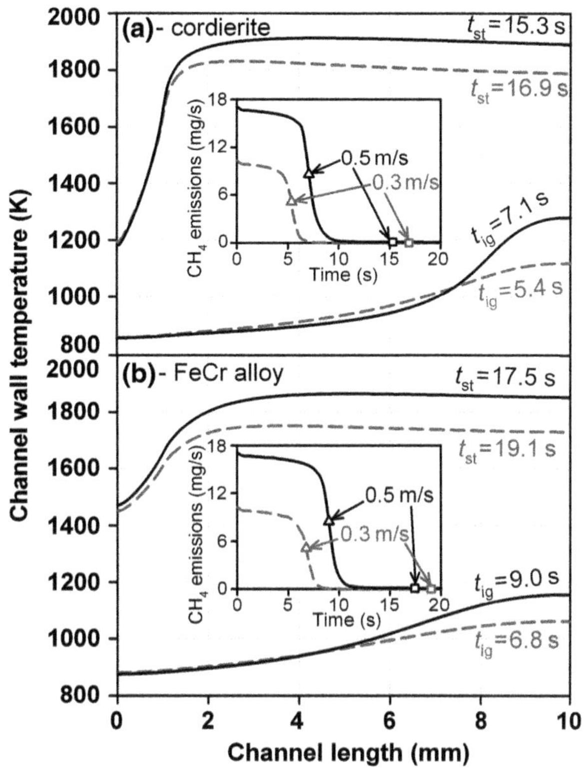

Fig. 8.13 Channel wall temperature profiles at t_{ig} and t_{st} for: (**a**) Cases 11 (*dashed lines*) and 13 (*solid lines*), (**b**) Cases 12 (*dashed lines*) and 14 (*solid lines*). *Solid lines*: $U_{IN} = 0.50$ m/s, *dashed lines*: $U_{IN} = 0.30$ m/s. *Inset*: methane emissions over the elapsed time until steady state. *Triangles*: t_{ig}. *Squares*: t_{st}

reactor period accounts for most of the unburned CH_4 exiting the channel. In the insets of Fig. 8.13, methane emissions are plotted versus time, considering a reactor width (z-direction) of 1 cm. By integrating over time until t_{st}, the cumulative emissions for Case 13 are more than double compared to those of Case 11 (118.5 versus 51.3 mg CH_4); the same holds true for Cases 14 and 12 (147.8 versus 64.3 mg CH_4, respectively).

8.3.5 Surface Radiation Heat Transfer

In numerical studies of catalytic microreactors and their large-scale counterparts, an energy transfer mechanism usually neglected is surface radiation exchange between channel wall elements and between the wall elements and the inlet/outlet sections. A number of numerical studies pointed towards the importance of radiation exchange in computational models when steep temperature gradients are present inside catalytic channels [17, 22, 23]. Moreover, it was shown in the foregoing Sect. 8.3.4 that radiation losses can be substantial in microreactor geometries. During the ignition and heat-up of microreactors with low-thermal-conductivity (mainly ceramic) walls, significant spatial temperature gradients can be created between the front- and rear-end sections (see, for example, Figs. 8.9 and 8.13).

In order to isolate the effect of the surface radiation heat transfer between reactor elements on t_{ig} and t_{st}, Cases 5 and 10 have been recomputed under adiabatic reactor conditions by suppressing both direct and reflective ($\varepsilon_{IN} = \varepsilon_{OUT} = 0.0$) radiation exchange with the inlet/outlet enclosures (allowing only for in-channel radiation exchange). Surface radiation of the channel walls was subsequently turned on or off by setting the channel surface emissivity either to its standard value ($\varepsilon = 0.6$) or to zero ($\varepsilon = 0.0$). Cases 15 and 16 in Table 8.2 pertain to a cordierite channel wall, while Cases 18 and 19 to a FeCr alloy one. Wall temperature profiles for Cases 15 and 16 at various time instances during the start-up phase are plotted in Fig. 8.14.

There are pronounced differences in the wall temperature profiles between calculations with and without surface radiation for cordierite microreactors, which increase as time progresses past ignition. The role of radiation can further be clarified with the aid of Fig. 8.15, where the local energy balance in the solid is presented at ignition for Case 16. All energy terms in the solid balance are plotted (heat generated due to surface reactions, heat convected to the gas, heat conducted in the solid, heat exchange via radiation, and accumulated heat), along with the wall temperature. Due to the hot spot created at the channel rear-end, a steep gradient ($\sim$ 300 K/cm) is noticeable along the channel. As a result, there is a net radiation heat transfer exchange between the wall elements, which is of the same order of magnitude as heat conduction in the solid.

By comparing t_{ig} and t_{st} for Cases 15 and 16 (Fig. 8.14), it is evident that radiation plays a dual role during start-up—at least for microreactors with low wall thermal conductivity. At ignition, radiation transfers heat away from the reaction zone (hot spot), reducing the wall temperature and thus increasing t_{ig} (see the negative Q_{RAD} term in Fig. 8.15, for $x > 8$ mm). However, following ignition, radiation helps in the faster redistribution and upstream transfer of heat inside the channel (see Fig. 8.14), thus reducing t_{st}; for Case 16, the reduction is up to $\sim$ 9.5%.

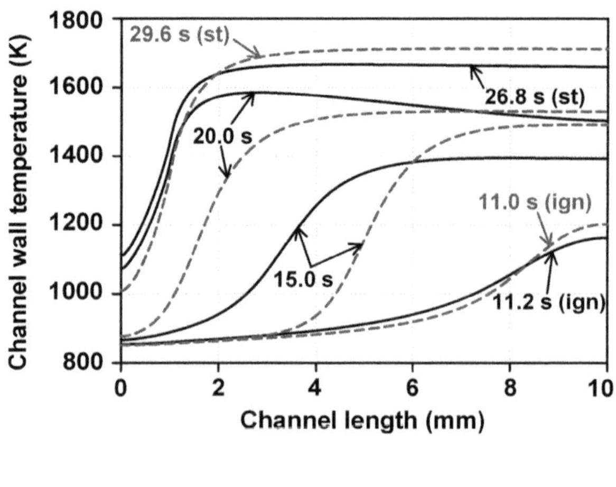

Fig. 8.14 Streamwise channel wall temperature profiles for Cases 15 (*dashed lines*, no surface radiation exchange) and 16 (*solid lines*, with in-channel surface radiation exchange, $\varepsilon = 0.6$) at four time instances, including ignition (ign) and steady state (st). Cordierite solid material

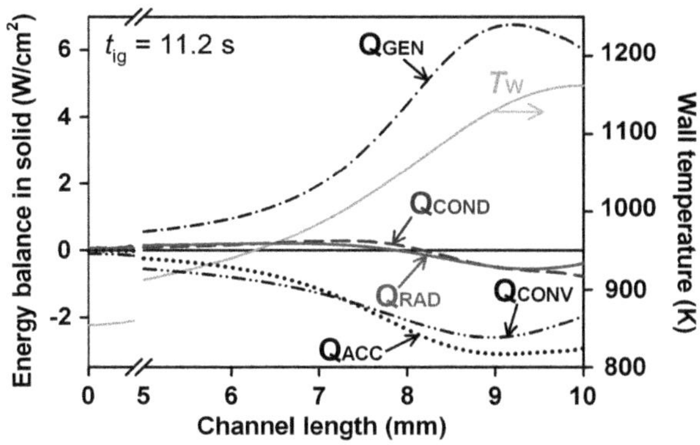

Fig. 8.15 Streamwise energy balance and temperature profile of the solid wall at ignition (t_{ig} = 11.2 s) for Case 16. *Dashed-dotted line*: heat generated via surface reactions (Q_{GEN}). *Dashed line*: heat conducted through the channel wall (Q_{COND}). *Solid line*: net radiation heat transfer (Q_{RAD}). *Dashed double-dotted line*: heat convected to the gas (Q_{CONV}). *Dotted line*: energy accumulated in the solid (Q_{ACC}). *Gray line*: channel wall temperature profile at t_{ig}. The rear 5 mm are shown in expanded detail

This behavior is further attested in Fig. 8.16, wherein the magnitudes of radiation and heat conduction in the solid are plotted during the heat-up phase. An abrupt change in the solid heat conduction profile is evident in all time instances at $x = 1$ mm; since the first 1 mm is catalytically inert, a step rise in the surface generated heat occurs at $x = 1$ mm, which affects the local energy balance by changing the slopes of Q_{COND} and Q_{CONV} at that point.

For $t_{ig} < t < t_{st}$, radiation becomes a major upstream heat transfer mechanism in the channel, enhancing the upstream propagation of the reaction front. A numerical experiment was conducted to quantify the effect of radiation on the "apparent" thermal conductivity of the solid wall, by recomputing Case 15 and assuming an artificial material (cordierite*) for the microreactor wall, which had essentially the same properties with cordierite in Table 8.1 except for a higher thermal conductivity; the value of k_s was increased at 0.5 W/mK steps, until t_{ig} and t_{st} assumed values close to the ones of Case 16. For $k_s = 5.5$ W/mK, ignition and steady-state times for a microreactor without surface radiation, became essentially identical to the characteristic times of a microreactor with $k_s = 2.0$ W/mK and a surface emissivity of $\varepsilon = 0.6$ (see Case 17 in Table 8.2). This apparent increase in the thermal conductivity of the solid, however, should not be confused with numerical models employing "effective" wall thermal conductivity for catalytic monoliths [23].

Finally, for microreactors with high wall thermal conductivity (Cases 18 and 19 of Table 8.2), simulations show that the wall temperatures become more spatially uniform and heat conduction in the wall dominates. This results in a weaker impact of radiation towards redistribution of energy within the channel for FeCr alloy microreactors (this is in contrast to the previously mentioned more pronounced effect of radiation on heat losses in the FeCr alloy).

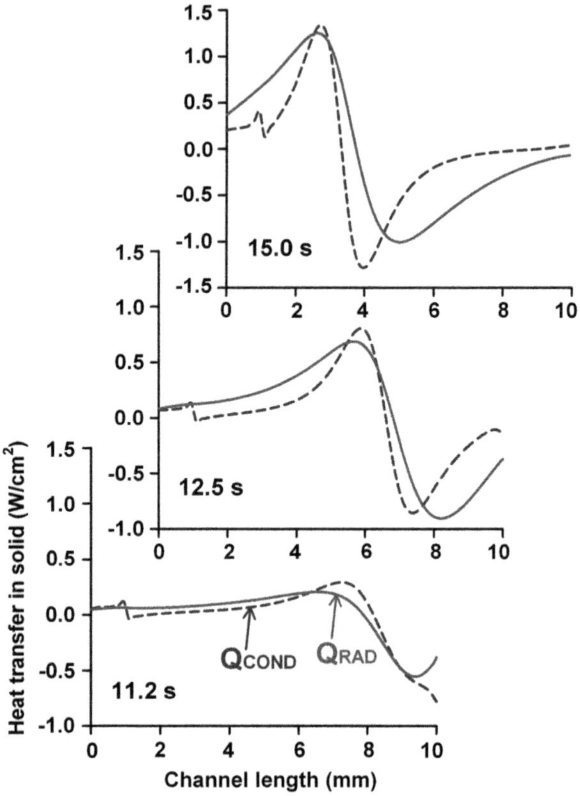

Fig. 8.16 Heat conduction in the solid and surface radiation in the microreactor wall for Case 16, at three times: $t = 11.2$ (ignition), 12.5 and 15.0 s. *Solid lines*: surface radiation. *Dashed lines*: solid heat conduction

8.3.6 Impact of Gas-Phase Chemistry

Gas-phase reactions are usually neglected in numerical investigations of catalytic microreactors. However, recent studies have pointed out to the importance of homogeneous chemistry in enhancing steady-state combustion stability against external heat losses, especially at elevated pressures ($p = 5$ bar) [17]. The enhanced combustion stability is not always a result of the total oxidation of methane via gas-phase reactions; an important coupled hetero-/homogeneous reaction route is the incomplete oxidation of methane to CO via gas-phase reactions, followed by the main exothermic oxidation of the formed CO to CO_2 not via homogeneous but via catalytic reactions.

In this work, a selected number of simulations were performed, wherein detailed gas-phase chemistry was included in the model. Cases 1, 5, 6 and 10–12 were recomputed with full hetero-/homogeneous chemistry; the t_{ig} and t_{st} for these simulations are reported in Table 8.2 as Cases 20–25, respectively. In Fig. 8.17, 2-D distributions are given for methane mass fraction and temperature for Case 5, and methane/OH mass fractions and temperature for Case 21.

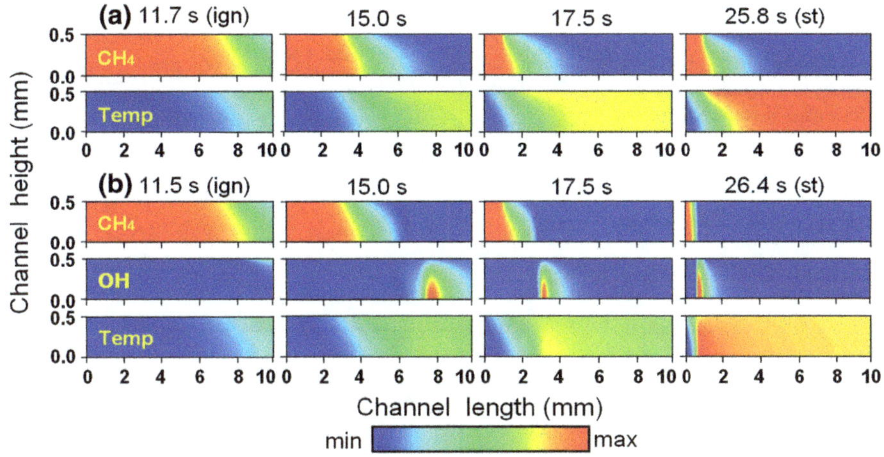

Fig. 8.17 Two-dimensional distributions of methane and OH radical mass fractions and of gas-phase temperature for (**a**) Case 5 and (**b**) Case 21, at four time instances, including ignition (ign) and steady state (st). The *color bar* defines minimum and maximum values. CH_4: 0.0–0.022; temperature: 850–1,708 K. OH radical: minimum 0.0 and maximum 0.59×10^{-6} for $t = 11.5$ s, 59.77×10^{-6} for $t = 15.0$ s, 259×10^{-6} for $t = 7.5$ s, and 766×10^{-6} for $t = 26.4$ s

Gas-phase chemistry affects ignition and steady-state times by contributing to the heat generation inside the channel and altering the spatial extent of the reaction zone over which fuel is consumed. At both investigated pressures of $p = 1$ and 5 bar and for both cordierite and FeCr alloy materials, gas-phase reactions only have a small impact on t_{ig}. In Cases 20 and 21 the ignition time is slightly reduced compared to Cases 1 and 5 by 0.3 and 0.2 s, respectively; similarly, for Cases 22 and 23 t_{ig} is reduced by 0.5 and 0.2 s compared to Cases 6 and 10, respectively.

Streamwise catalytic and gas-phase (the latter integrated over the y-direction) methane reaction rates and wall temperatures are plotted in Fig. 8.18 for Case 21 at 4 time instances; the reactor wall temperatures at ignition, steady state and two intermediate time instances for Case 5 are also plotted for comparison. At ignition, $t_{ig} = 11.5$ s, the weak exothermicity from homogeneous CH_4 conversion contributes to a slightly higher wall temperature (see top graph of Fig. 8.18) compared to the case with catalytic reactions only; this result is valid for both cordierite and FeCr alloy materials. Thus, surface reactions play the dominant role in determining microreactor ignition times.

It is also worth noting that, once at steady state, the microreactor of Case 21 operates essentially as a gas-phase microcombustor. This is evidenced by the sharp rise of gas-phase conversion near the channel entrance (t_{st} in Fig. 8.18) and the absence of significant catalytic conversion (the latter amounting to only 0.0045% of the former at $x > 1$ mm). The hot reactor walls facilitate in this case vigorous gas-phase reactions already at $x < 1$ mm. The large preheating of the incoming fuel/air mixture ($T_{IN} = 850$ K), the low inlet velocity U_{IN} and the high

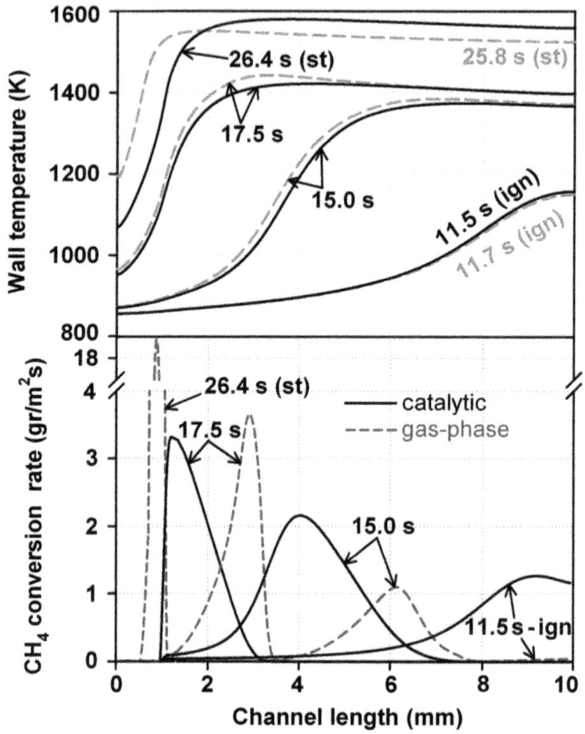

Fig. 8.18 Streamwise catalytic (*solid lines*) and gas-phase (*dashed lines*) methane conversion rates for Case 21 at four time instances (*bottom graph*) with the corresponding channel wall temperature profiles (*solid lines, top graph*). The channel wall temperature profiles at four time instances for Case 5 (catalytic chemistry only) are also shown on the top graph (*dashed lines*)

temperature of the reactor walls favor the formation of strong flames at the microreactor front-end.

In the higher equivalence ratio $\varphi = 0.6$, t_{ig} is insensitive to the presence of homogeneous reactions (compare characteristic times in Table 8.2 of Case 24 with 11 and Case 25 with 12). Since both catalytic and gas-phase reaction rates intensify as the fuel-to-air equivalence ratio increases, the competition for fuel consumption between the two chemical pathways becomes of prime importance. While the homogeneous reaction rate of methane has a $[CH_4]^{-0.3}$ dependence on methane concentration [24], the heterogeneous one has a $[CH_4]^{+1.0}$ dependence [9] and thus benefits to a greater extent as φ increases. A faster catalytic depletion of the fuel in the near-wall boundary layer diminishes the impact gas-phase reactions have on microreactor ignition times; characteristically, for Case 22 at $t_{ig} = 11.5$ s the homogeneous fuel conversion amounts to $\sim$ 3.6% of the heterogeneous one, while for Case 24 at $t_{ig} = 5.4$ s this percentage drops to $\sim$ 2.4%. It should be noted, however, that this behavior holds only during the early stages of microreactor ignition, since strong flames are also established for cases with $\varphi = 0.6$ as steady state is approached.

Following ignition and during the heat-up phase, the contribution of the homogeneous reaction pathway in fuel conversion increases substantially (see Fig. 8.18), with flames sustained in the channel by the exothermicity of the catalytic reactions, as also shown in Fig. 8.17b by the rising OH radical concentration while approaching steady state.

With a fraction of the heat release zone now shifted from the surface to the gas, the solid substrate is partly heated via convection and partly via direct heat generation on the surface. At a time instance ~ 17.5 s when the maximum value of gas-phase fuel conversion rate surpasses the corresponding catalytic one (see Fig. 8.18), the locus of maximum temperatures inside the microreactor shifts from the channel surface to the gas phase.

This shift is manifested by a high temperature core in the gas phase, seen in Fig. 8.17b at $t = 17.5$ s; the maximum reactor temperatures are further retained in the gas phase till steady state (see Fig. 8.17b at $t = 26.4$ s). Because the thermal conductivity of the hot gases is approximately two orders of magnitude lower than the thermal conductivity of cordierite, both heat accumulation in the wall and upstream propagation of the high temperature front zone of the solid are hindered by the additional thermal resistance of the gas phase, resulting in elongated characteristic steady-state times. The lower wall temperatures at the two time instances $t = 15$ and 17.5 s of Case 21 compared to those of Case 5 (Fig. 8.18) attest to this phenomenon.

In microreactors with FeCr alloy walls, t_{st} is impacted more severely by the larger difference between gas and solid thermal conductivities, with the time required for steady state increasing by as much as ~ 17.2% for Case 12 when gas-phase reactions are taken into account (compare with corresponding Case 25 in Table 8.2). Regardless of reactor material and operating conditions, the role of homogeneous chemistry in the start-up of methane-fueled catalytic microreactors is to elongate the steady-state times.

At $p = 1$ bar, the role of gas-phase chemistry in extending the steady-state times of FeCr alloy microreactors is diminished. By comparing t_{st} for Cases 6 and 22 ($p = 1$ bar, Table 8.2) an increase of 1.1 s (2.6%) is evident, while for Cases 10 and 23 ($p = 5$ bar, Table 8.2) the increase is 2.4 s (8.3%). Due to the reduced gas-phase reactivity of methane at lower pressures, the negative effect of gaseous chemistry in elongating the steady-state times is accordingly reduced.

On the other hand, cordierite microreactors exhibit an opposite behavior. Characteristically, while at $p = 1$ bar homogeneous chemistry extends t_{st} by 0.9 s (Cases 1 and 20), the corresponding extension at 5 bar is only 0.6 s (Cases 5 and 21). Despite the higher gas-phase reactivity of methane at $p = 5$ bar, the corresponding increase in catalytic reactivity with rising pressure results in a faster fuel depletion at wall temperatures lower than those at $p = 1$ bar. In the particular case of cordierite material, the higher catalytic reactivity leads to fuel conversion close to the mass transport limit (see Fig. 8.8), which in turn leaves significantly less amounts of methane available for homogeneous combustion.

Finally, microreactor CH_4 and CO emissions during the start-up phase are plotted for Cases 11 and 24 in Fig. 8.19, considering a channel width (z-direction) of 1 cm. When neglecting gas-phase chemistry, the amount of unburned CH_4 emissions is overpredicted, since homogeneous reactions further consume CH_4 (either via the direct gas-phase complete oxidation of CH_4, or via the

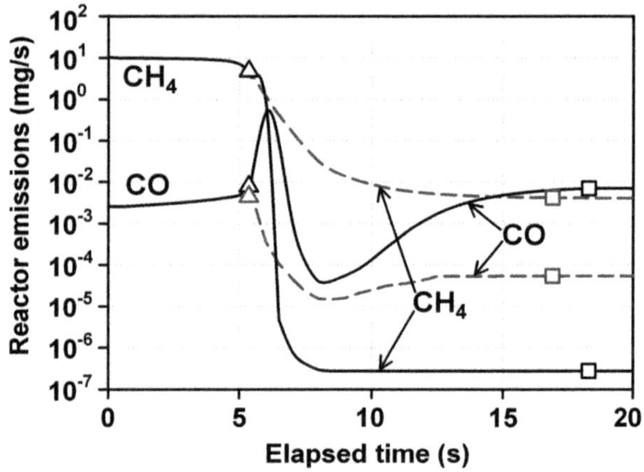

Fig. 8.19 Microreactor emissions of unburned CH_4 and CO versus elapsed time for Case 11 (catalytic reactions only; *dashed lines*) and Case 24 (catalytic and gas-phase reactions; *solid lines*). *Triangles*: ignition time t_{ig}. *Squares*: steady-state time t_{st}. A channel width of 1 cm is considered

homogeneous CH_4 to CO route coupled to the following catalytic route CO to CO_2) and do not allow for any significant fuel breakthrough during the heat-up phase.

On the other hand, CO emissions are predicted significantly higher when gas-phase chemistry is included, peaking around t_{ig}. The large activation energy for CO desorption from Pt makes the microreactor surface a poor source of this emission; however, incomplete combustion of CH_4 in the gas phase can contribute substantially to increased CO emissions even at steady state.

8.4 Conclusions

The start-up of methane-fueled, catalytic, channel-flow microreactors has been investigated numerically with a transient code that included full elliptic flow description, detailed hetero-/homogeneous chemistry, and all relevant heat transfer mechanisms in the reactor. Of particular interest were operating conditions pertinent to microturbine-based microreactor systems. Parametric studies have been carried out to identify the effect of various operations parameters, such a pressure, equivalence ratio, solid material properties and radiation properties on the transient process leading to ignition and finally to steady state operation. The following are the key conclusions of this study.

(1) Increasing the reactor operating pressure from 1 to 5 bar, while keeping the same mass throughput, substantially decreased both ignition (t_{ig}) and steady-state (t_{st}) times due to the positive pressure dependence of the catalytic reactivity, which in turn accelerated the formation and upstream propagation of the reaction zone. Both ignition and steady-state times were shorter for channels with ceramic cordierite material when compared to metallic FeCr alloy, since the lower thermal conductivity of the former allowed for the formation of a hot spot on the wall that facilitated ignition.

(2) Even though ignition and steady-state times where shorter in cordierite compared to FeCr alloy, the latter material has advantages for steady-state performance in terms of lower wall temperatures and wider steady combustion stability envelopes. Both opposing effects have to be considered when selecting appropriate reactor material, depending on the specific application.
(3) With catalytic reactivity playing the dominant role in microreactor ignition, an increase of the equivalence ratio resulted in a substantial decrease of both t_{ig} and t_{st} times of the start-up process. Increasing the inlet velocity also reduced the elapsed time until steady state was reached, owing to the higher input of chemical energy. However, it simultaneously increased the time required for reactor ignition and further deteriorated the cumulative reactor emissions (unburned CH_4).
(4) Surface radiation heat transfer played an important dual role for low thermal conductivity wall materials. On one side it increased t_{ig} by dissipating heat away from the initially formed hot spot zone. On the other side it reduced t_{st} due to a more effective redistribution of energy inside the channel, by transferring heat from the hot rear to the colder front section.
(5) Gas-phase chemistry played a minor role in the reactor ignition time, but had an appreciable effect in increasing the steady-state time, especially for high-thermal-conductivity microreactors, due to a less effective heat transfer from the reacting gas to the microreactor wall during the heat-up phase. Moreover, an increase in pressure from 1 to 5 bar augmented the impact of gas-phase chemistry on t_{st}, predominantly for the FeCr alloy microreactors.
(6) The presence of gas-phase reactions resulted in reduced CH_4 emissions at transient operation and steady state and in corresponding increased CO emissions. This suggests care when designing microreactor systems with models using only heterogeneous chemistry.

References

1. Schneider A, Mantzaras J, Eriksson S (2008) Ignition and extinction in catalytic partial oxidation of methane-oxygen mixtures with large H_2O and CO_2 dilution. Combust Sci Technol 180:89–126
2. Dogwiler U, Benz P, Mantzaras J (1999) Two-dimensional modeling for catalytically stabilized combustion of a lean methane-air mixture with elementary homogeneous and heterogeneous chemical reactions. Combust Flame 116:243–258
3. Schneider A, Mantzaras J, Jansohn P (2006) Experimental and numerical investigation of the catalytic partial oxidation of CH_4/O_2 mixtures diluted with H_2O and CO_2 in a short contact time reactor. Chem Eng Sci 61:4634–4646
4. Schneider B, Karagiannidis S, Bruderer M, Dyntar D, Zwyssig C, Guangchun Q, Diener M, Boulouchos K, Abhari RS, Guzzella L, Kolar JW (2005) Ultra-high-energy-density converter for portable power. Power-MEMS 2005, November 28–30, Tokyo, Japan
5. Isomura K, Murayama M, Teramoto S, Hikichi K, Endo Y, Togo S, Tanaks S (2006) Experimental verification of the feasibility of a 100 W class micro-scale gas turbine at an impeller diameter of 10 mm. J Micromech Microeng 16:S254–S261

6. Touloukian YS, Powel RW, Ho CY, Klemens PG (1970) Thermophysical properties of matter. Plenum, New York
7. Deutschmann O, Maier LI, Riedel U, Stroemman AH, Dibble RW (2000) Hydrogen assisted catalytic combustion of methane on platinum. Catal Today 59:141–150
8. Warnatz J, Dibble RW, Maas U (1996) Combustion, physical and chemical fundamentals, modeling and simulation. Springer-Verlag, New York
9. Reinke M, Mantzaras J, Schaeren R, Bombach R, Inauen A, Schenker S (2004) High-pressure catalytic combustion of methane over platinum: in situ experiments and detailed numerical predictions. Combust Flame 136:217–240
10. Reinke M, Mantzaras J, Bombach R, Schenker S, Inauen A (2005) Gas-phase chemistry in catalytic combustion of methane/air mixtures over platinum at pressures of 1 bar to 16 bar. Combust Flame 141:448–468
11. Kee RJ, Dixon-Lewis G, Warnatz J, Coltrin ME, Miller JA (1996) A fortran computer code package for the evaluation of gas-phase multicomponent transport properties. Report No. SAND86-8246; Sandia National Laboratories
12. Coltrin ME, Kee RJ, Rupley FM (1996) Surface Chemkin: a fortran package for analyzing heterogeneous chemical kinetics at the solid surface-gas phase interface. Report No. SAND90-8003C; Sandia National Laboratories
13. Kee RJ, Rupley FM, Miller JA (1996) Chemkin II: a fortran chemical kinetics package for the analysis of gas-phase chemical kinetics. Report No. SAND89-8009B; Sandia National Laboratories
14. Sinha N, Bruno C, Bracco FV (1985) Two-dimensional, transient catalytic combustion of CO-air on platinum. Physicochem Hydrod 6:373–391
15. Lutz AE, Kee RJ, Miller JA (1996) SENKIN: a fortran program for predicting homogeneous gas phase chemical kinetics with sensitivity analysis. Report No. SAND87-8248; Sandia National Laboratories, Report No. SAND87-8248
16. Deutschmann O, Schmidt R, Behrendt F, Warnatz J (1996) Numerical modeling of catalytic ignition. Proc Combust Inst 26:1747–1754
17. Karagiannidis S, Mantzaras J, Jackson G, Boulouchos K (2007) Hetero-/homogeneous combustion and stability maps in methane-fueled catalytic microreactors. Proc Combust Inst 31:3309–3317
18. Mantzaras J, Appel C (2002) Effects of finite rate heterogeneous kinetics on homogeneous ignition in catalytically stabilized channel-flow combustion. Combust Flame 130:336–351
19. Karim AM, Federici JA, Vlachos DG (2008) Portable power production from methanol in an integrated thermoelectric/microreactor system. J Power Sources 179:113–120
20. Karagiannidis S, Mantzaras J, Bombach R, Schenker S, Boulouchos K (2009) Experimental and numerical investigation of the hetero-/homogeneous combustion of lean propane/air mixtures over platinum. Proc Combust Inst 32:1947–1955
21. Kaisare NS, Stefanidis GD, Vlachos DG (2009) Comparison of ignition strategies for catalytic microburners. Proc Combust Inst 32:3027–3034
22. Boehman AL (1998) Radiation heat transfer in catalytic monoliths. AIChE 44:2745–2755
23. Lee ST, Aris R (1977) On the effects of radiative heat transfer in monoliths. Chem Eng Sci 32:827–837
24. Westbrook CK, Dryer FL (1981) Simplified reaction mechanisms for the oxidation of hydrocarbon fuels in flames. Combust Sci Technol 27:31–43

Chapter 9
Conclusions Summary: Outlook

9.1 Conclusions Summary

The present work investigated experimentally and numerically the lean hetero-/homogeneous combustion of propane on platinum as well as numerically the steady-state and transient performance of methane- and propane-fueled catalytic microreactors. The ultimate goal was the in-depth study of catalytic microscale combustion for portable power applications, with particular focus on conditions pertinent to micro-gas-turbines. The following constitute the major findings and conclusions of this work.

The pure heterogeneous and the coupled hetero-/homogeneous combustion of lean propane/air mixtures over platinum was investigated experimentally in an optically accessible, channel-flow reactor and a global-step catalytic reaction for propane was established for pressures 1 bar $\leq p \leq$ 7 bar. Numerical predictions using a detailed homogeneous reaction scheme coupled with the established catalytic reaction step reproduced the onset of gas-phase ignition at moderate pressures ($p \leq$ 5 bar). The overall performance of the employed hetero-/homogeneous reaction models verified their applicability in the design of propane-fueled catalytic microreactors.

A propane-fueled, catalytic, mesoscale combustor was investigated numerically and experimentally to assess its applicability for a portable, gas–turbine-based power generation system. After detailed parametric numerical studies of a single catalytic channel, a subscale model of the metallic catalytic combustor was constructed. Experimental testing verified the suitability of the proposed mesoscale combustor after meeting the required power output at the nominal mass throughput. A continuum model was used to simulate the two-dimensional temperature field of the monolith and allowed for a detailed description of the heat loss mechanisms in the monolith.

The combustion and heat transfer processes in a catalytic microreactor (length and height of 10 and 1 mm, respectively) coated with Pt and fed with fuel-lean

S. Karagiannidis, *Catalytic Microreactors for Portable Power Generation*,
Springer Theses, DOI: 10.1007/978-3-642-17668-5_9,

preheated CH_4/air mixtures were investigated numerically. A full-elliptic 2-D numerical model was employed that included detailed hetero-/homogeneous chemical reaction schemes and heat transfer mechanisms in the solid. Simulations at 1 and 5 bar have shown that the homogeneous reaction pathway could not be ignored at microreactor-relevant confinements and operating conditions. When considering the same mass throughput, the stable combustion envelope at 5 bar was substantially wider than at 1 bar due to the increased reactivities of both catalytic and gas-phase reaction pathways at elevated pressures. At k_s between 20 and 50 W/mK, a range that covers many metal and metallic/ceramic materials, the allowable heat losses reached their highest value. Surface radiation heat transfer has a dual role. For conditions far from extinction, surface radiation moderates the wall temperatures by providing a heat loss mechanism to the colder inlet. However, near the stability borders radiation can stabilize combustion by redistributing energy via transfer of heat from the hot rear to the colder front of the channel and aiding the catalyst light-off.

The combustion stability of propane- and methane-fueled, Pt-coated catalytic microreactors was investigated numerically in a microchannel configuration at 1 and 5 bar, wall thermal conductivities 2 and 16 W/mK, and channel half-heights $b = 0.5$ and 0.15 mm. Stability diagrams of critical heat transfer coefficient h versus inlet velocity U_{IN} were constructed for both fuels. Propane, similar to methane, substantially benefits from increased operating pressures, due to a positive $p^{+0.75}$ dependence of catalytic reactivity. The impact of gas-phase reactions in extending the blowout limits of propane is more pronounced than in methane, because of the higher gaseous reactivity of the former fuel. Numerical experiments revealed that, at higher inlet velocities, the transverse diffusive transport of fuel on the catalytic surface becomes dominant in determining reactor stability, an effect accentuated at elevated pressures ($p = 5$ bar). As the channel half-height decreases from $b = 0.5$ to 0.15 mm, however, the fuel transport properties become less important, due to the associated shorter characteristic mass diffusion time scales. Finally, increasing the solid thermal conductivity from 2 to 16 W/mK widens the stability limits for both propane and methane.

The start-up of methane-fueled, catalytic, channel-flow microreactors has been investigated numerically with a transient code that included full elliptic flow description, detailed hetero-/homogeneous chemistry, and all relevant heat transfer mechanisms in the reactor. Increasing the reactor operating pressure from 1 to 5 bar, while keeping the same mass throughput, substantially decreased both ignition (t_{ig}) and steady-state (t_{st}) times due to the positive pressure dependence of the catalytic reactivity, which in turn accelerated the formation and upstream propagation of the reaction zone. Both ignition and steady-state times were shorter for channels with ceramic cordierite material when compared to metallic FeCr alloy. Even though ignition and steady-state times where shorter in cordierite compared to FeCr alloy, the latter material had advantages for steady-state performance in terms of lower wall temperatures and wider steady combustion stability envelopes. With catalytic reactivity playing the dominant role in microreactor ignition, an increase of the equivalence ratio resulted in a substantial decrease of both t_{ig} and t_{st} times of the start-up process. Increasing the inlet velocity also reduced the elapsed time until steady state was reached, owing to the higher input of chemical energy. Surface radiation heat transfer played an important dual role for low thermal conductivity wall materials. On one side it increased t_{ig} by dissipating heat away from the initially formed hot spot zone. On the other side it reduced t_{st} due to a more effective redistribution of energy inside the channel, by transferring heat from the hot rear to the colder front section. Gas-phase chemistry had a minor role in reactor ignition, but an appreciable effect in increasing t_{st}, especially for metallic microreactors, due to a less effective heat transfer from the reacting gas-phase to the reactor wall during the heat-up phase.

9.2 Outlook

The lack of a detailed surface reaction mechanism for propane/air combustion on platinum necessitated the use of a global-step reaction model in this work. Secondary hetero-/homogeneous chemistry interactions are not taken into account this way (such as the homogeneous conversion of C_3H_8 to CO and the subsequent conversion of the latter to CO_2 on the catalytic surface); a closer-to-reality description of the in-channel combustion processes in catalytic microreactors is thus not possible. While the impact of such a simplification in cases where total oxidation of the fuel is required is minimal, this does not hold true in catalytic partial oxidation applications. Therein, a detailed surface chemistry description would be necessary.

Another benefit of developing a detailed surface reaction scheme for propane combustion on platinum would be the associated possibility of establishing a basis on which surface mechanisms for higher hydrocarbons could be developed, much like the H_2/CO surface reaction descriptions were a subset for CH_4 heterogeneous combustion and as CH_4 can be for C_3H_8.

Lean homogeneous combustion of propane in air was described with a detailed homogeneous reaction mechanism. The large number of species in this mechanism impacted heavily the computational cost of all numerical investigations in this work. Further efforts could focus on either validating other, less detailed, mechanisms that could reproduce results from the optically accessible catalytic reactor or on reducing the currently used mechanism to a more compact version that would yield good agreement with experiments.

Finally, the currently available 2-D full elliptic model has provided valuable insight on the fundamental physics of catalytic reactors. The established model could further be used to investigate both the steady-state and transient behavior of a large number of reactor setups, ranging from catalytic microreformers which would produce fuel for micro fuel cells in situ, to advanced combustor solutions for large-scale power generation systems such as the rich catalytic/lean burn concept.

Curriculum Vitae

Personal

Name:	Symeon Karagiannidis
Nationality:	Hellenic
Place/Date of Birth:	Thessaloniki, 20.12.1981

Education

2005–2010	Doctoral studies, Department of Mechanical and Process Engineering, Swiss Federal Institute of Technology Zurich, Switzerland
1999–2004	Department of Engineering and Management of Energy Resources, School of Engineering, University of Western Macedonia, Greece
1996–1999	2nd Lyceum of Thessaloniki
1993–1996	2nd Gymnasium of Thessaloniki
1986–1993	80th Elementary School of Thessaloniki

Working/Academic Experience

09.2009–08.2010	Paul Scherrer Institute, Villigen PSI, Switzerland Post Doctoral Fellow Technologies for gas turbine power generation with CO_2 mitigation

08.2005–05.2009	Paul Scherrer Institute, Villigen PSI, Switzerland Doctoral Researcher Experimental and numerical investigation of lean hetero-/homogeneous propane/air combustion on platinum and numerical studies of hydrocarbon-fueled catalytic micro-reactors for portable power generation
01.2005–08.2005	Aristotle University of Thessaloniki, Greece Research Assistant Numerical investigation of laminar jet, propane diffusion flame instabilities
11.2004–12.2004	Center for Research and Technology of Thessaloniki, Greece Research Assistant Catalytic pyrolysis of biomass for industrial chemicals production
07.2003–08.2003	Technical University of Bratislava, Slovakia Research Assistant Optimization and management of natural gas pipeline networks